50 VISIONS OF MATHEMATICS

Edited by

SAM PARC

Institute of Mathematics and its Applications

UNIVERSITY PRESS

OXFORD
UNIVERSITY PRESS

Great Clarendon Street, Oxford, OX2 6DP,
United Kingdom

Oxford University Press is a department of the University of Oxford.
It furthers the University's objective of excellence in research, scholarship,
and education by publishing worldwide. Oxford is a registered trade mark of
Oxford University Press in the UK and in certain other countries

Published in the United States of America by Oxford University Press
198 Madison Avenue, New York, NY 10016, United States of America

British Library Cataloguing in Publication Data
Data available

Library of Congress Control Number: 2013948420

ISBN 978–0–19–870181–1

Printed and bound by
CPI Group (UK) Ltd, Croydon, CR0 4YY

FOREWORD:
A BAD MATHEMATICIAN'S APOLOGY

DARA O BRIAIN

Whenever an anthology of great mathematics is compiled, the foreword, the warm-up man, is the best position I'll get on the bill. And this is for good reason. I started on the same path as these people, who will be here in a page or two with their bright ideas and their elegant proofs. In fact, if multiverse theory is to be believed there may even be a region of spacetime in which I continued there, struggling to keep up. This is our reality, though, and in this one, they kept going with the equations and I ran off to the circus. I left a field where we compare the properties of one thing to another in the hope of furthering our knowledge of both, and joined a trade where we say things like 'Sex is a lot like a bank account …'

This should never be construed as maths' loss for comedy's (arguable) gain.

Studying the mathematical sciences requires not just a quick mind, but a dedicated one. An entire syllabus can often be series of linked statements rolling on logically from the very first. If you miss a lecture, perhaps because college was also where you discovered telling jokes to crowds, partying, and the sort of exotic women who weren't in your secondary school, you can find yourself playing a year-long game of catch-up. You still attend the lectures but it becomes an exercise in the transcription of mathematics rather than the comprehension of it.

In the only facet of my life at the time which could have been described as monastic, I would sit at a desk in my lectures, scratching out my personal copy of the great theorems as the lecturer argued them out for me on the blackboard. Each lecture would cover about six A4 pages, furiously scribbled in the hope of being re-read at some later date when it might make sense in context. The high-water mark of this was a proof in second-year group theory for something called *Sylow's theorems*. The details of what Sylow theorised needn't concern you now. What is important was that it took four one-hour lectures to explain/dictate what is essentially a single result, scribbled furiously over 27 sheets of A4 paper. I looked it up online a minute ago and could make no sense of it there either.

I recently downloaded an Oxford University course on quantum mechanics. 27 lectures, recorded 'live', covering the sort of topics I had only seen before through a blur of a newly discovered social life. Here was my chance to put right the mistakes of a younger man, sat in my home with a notepad and pen.

I pressed 'play' on the first one, the lecturer began, and immediately I had to start writing frantically as the blackboard was filled with line after line of symbols. For Proust, nostalgia was triggered by the taste of madeleine cake; for me, it was scribbling maths furiously and trying to keep up with a sustained argument. Luckily, just when I could feel it slipping away from me, the cat walked across the keyboard, stood on the space bar, and the video stopped. I was suddenly back to being a 41-year-old man, sitting in his kitchen on the family computer, but almost out of breath from the writing.

By contrast, a friend sneaked me into a third-year philosophy lecture once. Pen at the ready, notepad open, I found myself attending what I could only describe as 'a chat'. A man burbled amiably at the head of the class and the students nodded, deigning occasionally to jot the odd note down. The lecture was about the veil of perception, I recall, the theory that what we 'sense' about the reality around us is only what is fed through our physical senses and since this 'veil' exists between us and presumed, actual reality, we cannot come to any definite conclusions about the nature of that reality.

I might have that wrong. I almost definitely have some part of it incorrect. I'd even welcome it being corrected by any passing philosophy professor. The point is, though, that's what I remember from a single lecture 19 years ago. I did four lectures about Sylow's theorems, plus surrounding material and question sheets, and then an exam, which I presumably passed. And I still had to google it this evening to remind myself what it was about.

So, maths is intense. But you stick with that because it is so beautiful. And this beauty is often not clear to outsiders.

You'll often hear mathematicians talk about beauty. Usually it is in reference to an elegance or wit in the argument; an efficiency in paring a question down to its most important parts; how a few lines of simple logic reveal a deep truth about numbers, or shape, or symmetry.

There are other things that make maths beautiful, however, particularly in comparison to the other sciences. For a start, it is delightfully uncluttered with Latin names. Not for maths the waste of energy learning off genus, phylum, and kingdom of a snail that looks pretty much the same as the next snail, except for a slightly stripier shell. Or the proper name of that tube that connects the inner ear of a horse to that other part of the inner ear of the horse.

Equally, maths was never too bothered about measuring things. I spent four years doing maths and probably never saw a number, other than the one next to the question, or at the bottom of the page. We'll do the variables and somebody with an oscilloscope can plug in the numbers. Maths will predict the results and somebody else can go off and build a 17-mile-long tunnel under Switzerland and tell us if we're right. This is not just disdain; there's an admission of practical ineptitude implied too. There are 500,000 rivets in the Large Hadron Collider. If they'd left them to the mathematicians those packets of protons would have gone four feet before hitting a stray lump of solder. Or a cufflink. Or a mathematician soldered to the tube, by his cufflink.

I have become publicly associated with astronomy, for example, by virtue of a television show I do. However, I know I would be of no use as an actual astronomer, a field built on meticulous observation and measurement: tracing points of lights across the sky, night after night, in the hope of spotting the tiniest aberrations, revealing the hugest things. Or you could write down the equations of cosmology and let somebody else do all that. That's the beauty of maths.

Sometimes beauty is not enough, though.

I visited my head of department at Mathematical Physics at the end of my degree to drop the bombshell that I wouldn't be seeking a Masters place; this was so that he could politely feign surprise and disappointment. 'We'll miss you,' he said, rising to the challenge. 'You had a tremendous … flair … in your mathematics.' It was only after I'd left the room that I realised that this was not intended as a compliment. They had been looking for even a trace of rigour.

The pieces in this collection have plenty of flair. They sparkle with bright ideas and clever results. They're best read with a pen and paper so that you can try out some of the leaps in imagination yourself. There will often be the surprising results of comparing the properties of one thing to another. There may even be the odd good joke.

And if you ever feel intimidated during these pages, as some people do with maths, let me offer a reassurance.

At some point during their college career, each fine mind whose work is contained within this book was just another student furiously transcribing superscript i's and subscript j's, and hoping that they would get a moment's pause to catch their breath and pick the misplaced solder from their hands.

PREFACE

There is an old saying, that God made the whole numbers and that all of the rest of mathematics is the invention of man.* If that is true, what was going on when fifty was made? Because, from a human perspective, fifty seems to have a particular significance. We celebrate fiftieth birthdays, wedding anniversaries, and any other commemorations you might care to mention, with special verve. Fifty also represents a coming of age for many of us personally, an age where we might want to take on a fresh challenge, embrace new visions of life. The phrase 'fifty-fifty' implies a perfect degree of fairness. And fifty per cent is halfway there – a sense of achievement, yet hinting at more to come.

This book celebrates not just the number 50 and the concept of fiftiness, but the whole of the rest of mathematics, as well as the people involved in its creation, and its underpinning of many things we take for granted. It pays tribute to the relevance that mathematics has to all of our lives and embraces an exciting vision of the next 50 years of mathematics and its applications.

Fifty is of course also a number, and as a number it has its own special properties. To a mathematician, 50 is the smallest number that is the sum of two non-zero square numbers in two distinct ways, $50 = 1^2 + 7^2 = 5^2 + 5^2$; it is the sum of three successive square numbers, $50 = 3^2 + 4^2 + 5^2$. Moreover, to chemists it is a magic number, as atomic nuclei with 50 nucleons are especially stable.

The original motivation for this book was the 50th anniversary of the UK's Institute of Mathematics and its Applications (IMA) in 2014, but the project grew in scope. Moreover, it would seem that, during the last 50 years, mathematics has come of age. School mathematics has undergone a transition from arithmetic, algebra, and geometry alone to sets, topology, computer imagery, and an appreciation of the importance of mathematics in its many diverse applications. We have seen the birth of new scientific paradigms, for example in chaos theory, string theory, and genomics. We have experienced the computer revolution: nearly every home now has a computer with more calculating power than all of the world's computers combined 50 years ago. We have the new mathematics of the digital age, where massive datasets are pervasive and the mathematics-based technologies of the Internet have completely transformed our lives. We are also in the age of computer modelling: the entire human physiome and the climate of the Earth over millennia can now be simulated using billions of variables. Given this progress in the last 50 years, we must ask with a sense of wonder what the future of mathematics and its applications will be.

The main content of this book is a collection of 50 original essays contributed by a wide variety of authors. The topics covered are deliberately diverse and involve concepts from simple numerology to the very cutting edge of mathematics research. Each article is designed to be read in one sitting and to be accessible to a general audience. Nevertheless, I have not asked the contributors to shy away from using equations and mathematical notation where necessary. For those of little or no mathematical training, please don't worry; it should be possible to get the sense of any article without reading the symbols.

---

*This is attributed to the 19th-century Prussian mathematician Leopold Kronecker.

Contributors were chosen in a rather haphazard way. Aided by my editorial team (see below), we brainstormed a list of potential leading mathematicians who would like to share their vision of mathematics with us. We were delighted and humbled by just how many agreed, which left us with a difficult problem of selecting the final 50. In a few cases the piece is based on something that has appeared elsewhere in another form, notably in the IMA's *Mathematics Today* magazine or in the truly excellent *Plus Magazine* online (<http://plus.maths.org>). We also ran a competition in both *Mathematics Today* and *Plus* to invite new authors to contribute to this project. You will find a small selection of the best entries we received sprinkled among those from more established authors.

The essays are reproduced in alphabetical author order, which makes a somewhat random order thematically. Nevertheless, the contributors were asked for pieces that fall into one of five broadly defined categories: mathematics or mathematicians from the last 50 years (a time frame that is somewhat stretched in a couple of cases); quirky mathematics; mathematics of recreation; mathematics at work; and the philosophy or pedagogy of mathematics. Each piece is intended as a mere appetiser and, where appropriate, concludes with a few items of suggested further reading for those tempted into further delights.

There is also other content. There are 50 pictorial 'visions of mathematics', which were supplied in response to an open call for contributions from IMA members, *Plus* readers, and the worldwide mathematics community. Again, these images are presented in no particular order; nor are they necessarily supposed to be the top 50 images of mathematics in any objective sense. Mathematics is a highly visual subject, and these images are there to illustrate the breadth of that vision. An attribution and a very short description of each image are supplied at the back of the book. I have also been tempted to include other mathematical goodies related to the number 50. In particular, as I said at the start, $50 = 3^2 + 4^2 + 5^2$. It might not have escaped your attention that, in addition, $3^2 + 4^2 = 5^2$. This makes (3; 4; 5) the first *Pythagorean triple*, the smallest whole numbers for which there is a right-angled triangle with these side lengths: 3, 4, and 5. This (admittedly slender) link between 50 and Pythagoras's theorem allows me to introduce a running theme through the book, presented in three chunks – a series of 'proofs' of this immortal theorem in a number of different literary styles. I hope that you (and Raymond Queneau, from whose work my inspiration came) will forgive me for this.

There are many people who deserve thanks for their involvement in putting this book together. First and foremost, I should like to thank the IMA – particularly David Youdan, Robert MacKay, John Meeson, and members of the IMA council – for keeping the faith when what was originally envisaged as a small pamphlet to accompany their 50th birthday celebrations grew and GREW. All the profits from the sale of this book will go to support the current and future work of the IMA in promoting and professionalising mathematics and its applications.

I am deeply indebted to all the contributors to this book, who have fully embraced the spirit of the project, and have given their work for free, agreeing to waive any royalties. Also, particular thanks go to Steve Humble, Ahmer Wadee, Marianne Freiberger, Paul Glendinning, Alan Champneys, Rachel Thomas, and Chris Budd, the informal editorial committee of IMA members who have aided me in all stages of this project, from suggesting contributors and liaising with them, to editing the copy. Keith Mansfield at OUP has been invaluable and has gone way beyond what one might normally expect of a publishing editor. I should like to single out Maurice MacSweeney, without whose forensic skills key parts of the book's manuscript might have been lost forever. I should acknowledge Nicolas Bourbaki, whose style of writing has greatly influenced the way I work. And, finally, thanks go to my family and to Benji the dog, who have had to suffer fifty forms of hell as I have juggled the requirements of my many other commitments.

CONTENTS

CHAPTER 1

What's the problem with mathematics?

DAVID ACHESON

W hy do so many people have some kind of problem with mathematics?
The real truth, so far as I can see, is that most of them are never let anywhere near it.
They see mathematics as being about aimless calculations, rather than about discovery and adventure. In particular, they see none of the surprise that often comes with mathematics at its best. Yet I had my first big mathematical surprise at the age of just 10, in 1956. I was keen on conjuring at the time, and came across the following mind-reading trick in a book – see Fig. 1.1.

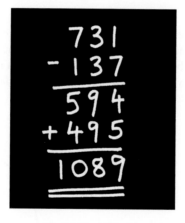

Fig 1.1 The 1089 trick.

Think of a three-figure number. Any such number will do, as long as the first and last figures differ by 2 or more. Now reverse your number, and subtract the smaller three-figure number from the larger. Finally, reverse the result of that calculation and add. Then the final answer will always be 1089, no matter which number you start with! And while it may not be very serious mathematics, I have to tell you this: if you first see it as a 10-year-old boy in 1956, it blows your socks off (see Chapter 34 for an even more subtle number puzzle).

Start with geometry?

Over many years now I have tried to share my sense of wonderment with mathematics through so-called family or community lectures in schools, usually held in the evening. The age range at such events can be enormous, from grandparents to very young children indeed. And all you can really assume is that each family group has at least one person who is good at sums.

Now the fastest way I know of introducing the whole spirit of mathematics at its best, especially to a very young person, is through geometry. And, as it happens, I once had the opportunity to try this out, with the help of a teacher, on a small group of 8-year-olds in a primary school in Hertfordshire. We started with parallel lines, and used these to prove that the angles of a triangle always add up to 180°. You can see this by staring hard at Fig. 1.2.

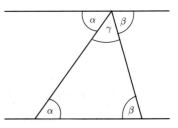

Fig 1.2 A geometric proof that the angles in a triangle sum to 180°

Next, we noted that the base angles of an isosceles triangle are equal, and agreed that this was fairly obvious. (The clock was ticking.) Then they all did some practical experimentation, finding that the angle in a semicircle always seems to be 90°. This caused a real stir, and one of them even shrieked in amazement. Finally, we used our two results on triangles to prove this, as reproduced in Fig. 1.3.

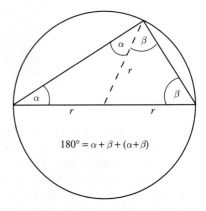

$$180° = \alpha + \beta + (\alpha + \beta)$$

Fig 1.3 Proof that the angle inscribed in a semicircle is 90°.

As I recall, we made this particular journey – from almost nothing to what is arguably the first great theorem in geometry – in just half an hour or so. And nobody burst into tears.

Proof by pizza

With a slightly older group, the whole idea of an infinite series, such as

$$1/4 + 1/16 + 1/64 + \ldots = 1/3,$$

offers some real possibilities. Many people are genuinely surprised that such a series can have a finite sum at all. And, in my experience, they can be even more struck by the elegance of an off-beat derivation of this result, which I call *proof by pizza*.

Take a square pizza, of side length 1 unit (a foot, say), which therefore has area 1. Cut it into four equal pieces, and arrange three of these in a column, as in Fig. 1.4. Then cut the piece that is left over into four equal pieces, and arrange three of those, likewise, in a column as in panel (b). Now keep on doing this for ever. In this way (if you ignore bits of cheese falling off, etc.), you generate three identical rows. Each row is, in terms of area, equivalent to the original infinite series. The total area must still be 1, so each series must add up to 1/3. Then you eat the pizza (c), and that completes the proof.

Fig 1.4 Proof by pizza.

It is then, arguably, a smallish step to present – albeit without proof – one of the more subtle pleasures of mathematics at its best, namely *unexpected connections* between different parts of the subject. For this purpose, the so-called Gregory–Leibniz series (which was actually first discovered in India) shown in Fig. 1.5 is, in my view, hard to beat. After all, everybody knows what an odd number is, and everybody knows that π is all about circles. But why should these two ideas be related at all, let alone in this beautifully simple way?

$$\frac{\pi}{4} = 1 - \frac{1}{3} + \frac{1}{5} - \frac{1}{7} + \ldots$$

Fig 1.5 The Gregory–Leibniz series.

Not quite the Indian rope trick

One of the major functions of mathematics, surely, is to help us understand the way the world works, and, in particular, to get where physical intuition cannot reach. I like to share with audiences my most memorable experience of this, which came one wet, windy afternoon in November 1992. For some weeks before, strange things had been happening in my computer models of multiple pendulums, so I finally sat down with a blank sheet of paper and tried to find, and prove, a general theorem.

And just 45 minutes later, against all my intuition, it dropped out, and implied that a whole chain of *N* linked pendulums can be stabilised *upside down*, defying gravity – a little bit like the Indian rope trick – provided that the pivot at the bottom is vibrated up and down by a small enough amount and at a high enough frequency. You can see a picture of the pendulums in Fig. 1.6 on the board just behind the guitar-playing penguin, who I think is supposed to be me.

Fig 1.6 Steve Bell's interpretation of my room in Oxford.

And when my colleague Tom Mullin verified these predictions experimentally, the whole business rather captured the public imagination and eventually featured in newspapers and on national television. Now, the theorem is too quirky, I think, to be of any great significance for the future of the world, but it was still the most exciting 45 minutes of my mathematical life, and whenever I talk of that afternoon to young people I like to think that it spurs them on a bit.

Real or imaginary?

A few years ago, I wrote a book on mathematics for the general public, and the final chapter – on so-called imaginary numbers – started with the cartoon in Fig. 1.7. The equation on the television screen, due to Euler, is one of the most famous in the whole of mathematics, for it provides an extraordinary connection between π, the number $e = 2.7182818\ldots$ (the base of natural logarithms) and i, which is the 'simplest' imaginary number of all, namely the square root of -1.

Fig 1.7 Euler's formula.

With the television cartoon, I am effectively asking whether the above scene might really occur at some point in the future. Not perhaps with that particular family, or even with that particular television, but . . .

Some will say that this is hoping for too much, and I suspect the Oxford mathematician Edward Titchmarsh would have agreed, because he wrote, in his 1959 classic *Mathematics for the general reader*, 'I met a man once who told me that, far from believing in the square root of minus one, he didn't even believe in minus one.'

But I am an optimist. On one occasion, during one of my community lectures at a school in North London, I was midway through proof by pizza when I happened to notice a particular little boy, aged about 10, in the audience. And a split second after delivering the punchline of my proof, when a deep idea suddenly becomes almost obvious, I actually saw the 'light bulb' go on in his head, and he got so excited that he fell off his chair.

And, in a sense, that fleeting moment says it all.

· ·

FURTHER READING

[1] David Acheson (2010). *1089 and all that: A journey into mathematics*. Oxford University Press (paperback edition).
[2] David Acheson and Tom Mullin (1993). Upside-down pendulums. *Nature*, vol. 366, pp. 215–216.
[3] Edward Titchmarsh (1981). *Mathematics for the general reader*. Dover Publications.

CHAPTER 2

The mathematics of messages

ALAN J. AW

We all write messages, be they SMSs to our loved ones, emails to friends, or even telegrams. Yet surprisingly few of us ask, 'How are these messages or data, which are stored in the deep recesses of iCloud, the Internet repository, or some obscure "geek facility", transmitted with such clarity and speed?' Perhaps with a little help from physics, we could surmise that messages are transmitted by waves. However, this alone does not explain how the data are transmitted with high accuracy. In fact, we intuit that waves travelling along a non-vacuum medium are most likely to experience disturbances, or *perturbations*, which would introduce errors into the data being transmitted. Moreover, these perturbations are likely to be irreversible, i.e. the waves do not undergo a self-correcting mechanism. So, the principal question to ask is: how could data still be transmitted with such high fidelity? Or, in more pragmatic language, how is it that we enjoy such speedy and accurate means of communications every day?

Messages, information, and data

It turns out that there is an underlying mathematical theory behind all these data or *information* transfers. The transmission of data via waves from the Internet to our smart devices is a specific example of a more general and abstract notion of data transfer from one point to another. Here, a point is either a sender (i.e. an information source) or a receiver (i.e. a destination), and could be, for example, a satellite or a mobile phone.

In this general model of information transfer the sender first sends some information or message to an *encoder*, which then encodes it by simply representing the message using a suitable mathematical structure. A historically significant example is the use of binary digits or *bits*, i.e. zeros and ones, to encode black and white pictures. In this encoding technique, which was implemented by NASA in the 1960s, the picture was divided into equally sized boxes so that each box was either fully black or white; and the encoder used the digit 1 to represent every black box and 0 for every white box, effectively giving rise to an array of 1s and 0s (or, in mathematical parlance, an *incidence matrix*).

The next step after encoding is to communicate the encoded information, which we refer to as *data*, to a receiver. In doing so, the data are transmitted across a medium or *channel* – in the case of the satellite it would be the atmosphere and the areas of the galaxy near the surface of the

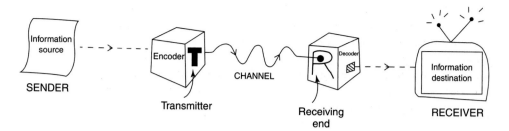

Fig 2.1 Information transfer.

Earth – to the receiving end. Finally, the receiving end *decodes* the data, that is, does exactly the opposite of what the encoder did to the information. *Et voilà*, the receiver obtains the original information. Figure 2.1 summarily illustrates the processes just described.

Surely the system of message transmission must be more complicated, for if not we would certainly start to question, indeed very indignantly, what mathematicians or engineers working in this subject area do exactly to earn their salary. There are two inherent problems present in the model above. First, as you might have already figured out from the example of wave perturbation, the channel introduces disturbances or *noise* into the data. This affects the accuracy or *reliability* of the data being transferred (see A in Fig. 2.2). Second, with reference to the NASA example, sometimes it is impossible to represent a picture, or in fact any message in general, by a mathematical structure 100% accurately. And this gives rise to what we call *information distortion*. (See B in Fig. 2.2.)

Have these problems been solved or addressed well? The answer is yes, but not quite yet (tricky indeed!). With regard to the first problem, also called the *channel coding problem*, mathematicians have figured that by adding extra elements that hold no meaning whatsoever in relation to the data being transferred (aptly termed *redundancies* in mathematical jargon), one can decrease the chance of noise affecting the fidelity of the data. The redundancies render the data less susceptible to the irreversible or permanent perturbations that make the data less accurate. However, this would slow down the speed or *rate* of data transmission, since the transmitter has to send additional redundancies across the channel at each moment. Eventually a compromise has to be made, but what is the best compromise we can achieve?

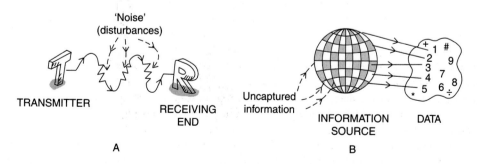

Fig 2.2 Impaired accuracy (A) and information distortion (B).

Let us keep this question in mind while we consider the second problem (the *source coding problem*). In representing our message by a mathematical structure, we have to use a set of symbols to encapsulate its various distinct elements. Patently we wish to capture the information fully without introducing distortions, but that would necessitate the usage of more symbols. Engineers worry about this because their aim is to compress the information as much as possible by using as few symbols as possible. Thus, a compromise has to be made, and the question of choosing the best compromise surfaces once more.

Shannon's theory of communication

In 1948 the mathematician Claude Elwood Shannon published two papers, collectively entitled 'A mathematical theory of communication', which described and analysed a general mathematical model of information transfer – indeed the one which we have described. Shannon proved that there are fundamental limits to both the rate of data transmission and the extent of information compression. That is to say, (i) beyond a certain rate, data transmission inevitably becomes unreliable; and (ii) below a certain level of data compression (i.e. usage of as few symbols as possible), there is bound to be information distortion.

Now these discoveries may seem obvious at first glance, but if we are a little pernickety we realise how charming they actually are. Delving first into a little mathematical detail, Shannon employed ideas from probability theory. He modelled the sender and transmitter each by a *random variable* effectively producing messages element by element, each time producing a certain element with a particular chance. Next, he ingeniously formulated a mathematical measure for the amount of information contained in a message, which he called *entropy*. In other words, if we denote a transmitter by the random variable X, then there is a function H which when applied to X gives $H(X)$, which is known as the entropy of X.

This $H(X)$ turns out to possess very powerful properties; hence, Shannon established the following interesting facts.

- $H(X)$ is the measure of the limit of data compression of an information source X without subjecting it to inevitable distortion. In other words, the higher the entropy, or information content, of your message, the less you can compress it.

- The ideas of the H function can be extended elegantly to arrive at a mathematical expression called *mutual information*. This turns out, somewhat unexpectedly, to be the measure of the limit of the reliable data transmission rate across a channel. In other words, the higher the mutual information, the greater the maximum rate of reliable data transmission.

Additionally, whereas beyond the calculated value of mutual information data transmission indubitably becomes unreliable, it is also true that at rates below this value arbitrarily reliable levels of communication are achievable. That is, one may choose a specific degree of error (caused by noise) allowable during data transmission, and there is always a corresponding rate below the mutual information which allows errors precisely up to that degree. Similarly, at any level of compression above the entropy, arbitrarily small degrees of information distortion can be attained. Perhaps all these go against our intuition that the unreliability or amount of distortion is a continuous relation of the rate or level of compression; for there are in reality very precise thresholds for both compressions and rates, anything beyond which leads to disaster.

Are our problems solved?

In view of Shannon's intellectual feat, it might seem that our two fundamental problems of communication have been addressed completely. Unfortunately, we are far from solving them. Unbeknown to most of us, mathematicians and engineers are actively and persistently figuring out ways to achieve the compression and rate limits – indeed, it is one thing to know the fundamental limits and another actually to attain them, and the latter is often the more challenging. At the same time, mathematicians often contemplate new ways of utilising their multitude of abstract structures to represent messages or information. In short, there is much unfinished business for the mathematics and engineering communities.

Admittedly, not many of us may be able to appreciate Shannon's brilliant ideas as being on the same levels as those of mathematicians and engineers. (Perhaps one would be mildly surprised to know that he is remembered as the father of information theory.) Nonetheless, at least we now know that thanks to Shannon theory, we are able to communicate so efficiently and effectively in a world riding increasingly swiftly and ineluctably on the tides of globalisation.

. . . .

FURTHER READING

[1] Claude Shannon (1948). A mathematical theory of communication. *Bell Systems Technical Journal*, vol. 27, pp. 379–423. Available for free at: <http://www.alcatel-lucent.com/bstj/vol27-1948/articles/bstj27-3-379.pdf>.

[2] Raymond Hill (1990). *A first course in coding theory*. Oxford Applied Mathematics and Computing Science Series: Oxford University Press, New York.

[3] James Gleick (2011). *The information: A history, a theory, a flood*. HarperCollins.

Source
Selected as an outstanding entry for the competition run by *Plus Magazine* (<http://plus.maths.org>).

CHAPTER 3

Decathlon: The art of scoring points

JOHN D. BARROW

The decathlon consists of ten track and field events spread over two days. It is the most physically demanding event for athletes. On day one, the 100 metres, long jump, shot put, high jump, and 400 metres are contested. On day two, the competitors face the 110 metre hurdles, discus, pole vault, javelin, and, finally, the 1500 metres. In order to combine the results of these very different events – some give times and some give distances – a points system has been developed. Each performance is awarded a predetermined number of points according to a set of performance tables. These are added, event by event, and the winner is the athlete with the highest points total after ten events.

The most striking thing about the decathlon is that the tables giving the number of points awarded for different performances are rather free inventions. Someone decided them back in 1912 and they have subsequently been updated on different occasions. Clearly, working out the fairest points allocation for any running, jumping, or throwing performance is crucial and defines the whole nature of the event very sensitively. Britain's Daley Thompson missed breaking the decathlon world record by one point when he won the Olympic Games, 1984, but a revision of the scoring tables the following year increased his score slightly and he became the new world record holder retrospectively!

All of this suggests some important questions that bring mathematics into play. What would happen if the points tables were changed? What events repay your training investment with the greatest points payoff? And what sort of athlete is going to do best in the decathlon – a runner, a thrower, or a jumper?

The decathlon events fall into two categories: running events, where the aim is to record the least possible time, and throwing or jumping events, where the aim is to record the greatest possible distance. The simplest way of scoring this would be to multiply all the throw and jump distances in metres together, then multiply all the running times in seconds together, and divide the product of the throws and jumps by the product of the running times. The *Special Total, ST,* would be

$$ST = \frac{LJ \times HJ \times PV \times JT \times DT \times SP}{T_{100\text{ m}} \times T_{400\text{ m}} \times T_{110\text{ mH}} \times T_{1500\text{ m}}}$$

where *LJ* is the distance from the long jump, *HJ* that from the high jump, *PV* that from the pole vault, *JT* that from the javelin, *DT* that from the discus, and *SP* that from the shot put, with units of

$$\frac{(\text{length})^6}{(\text{time})^4} = \frac{m^6}{s^4}.$$

If we take the three best-ever decathlon performances by Ashton Eaton (9039 points), Roman Šebrle (9026 points), and Tomáš Dvořák (8994 points), and work out the Special Totals for the ten performances they each produced then we get

Šebrle (9026 pts): $ST = 2.29$,
Dvořák (8994 pts): $ST = 2.40$,
Eaton (9039 pts): $ST = 1.92$.

Interestingly, we see that the second best performance by Dvořák becomes the best using this new scoring system, and Eaton's drops to third.

In fact, our new scoring system contains some biases. Since the distances attained and the times recorded are different for the various events, you can make a bigger change to the ST score for the same effort. An improvement in the 100 metres from 10.6 seconds to 10.5 seconds requires considerable effort but you don't get much of a reward for it in the ST score. By contrast, reducing a slow 1500 metre run by 10 seconds has a big impact. The events with the room for larger changes have bigger effects on the total.

The setting of the points tables that are used in practice is a technical business that has evolved over a long period of time. It pays attention to world records, the standards of the top-ranked athletes, and historical decathlon performances. However, ultimately it is a human choice, and if a different choice was made then different points would be received for the same athletic performances, and the medallists in the Olympic Games might be different. The 2001 IAAF scoring tables have the following simple mathematical structure.

The points awarded (decimals are rounded to the nearest whole number to avoid fractional points) in each track event – where you want to give higher points for shorter times – are given by the formula

$$A \times (B - T)^C,$$

where T is the time recorded by the athlete in a track event and A, B, and C are numbers chosen for each event so as to calibrate the points awarded in an equitable way. The quantity B gives the cut-off time at and above which you will score zero points. T is always less than B in practice – unless someone falls over and crawls to the finish!

For the jumps and throws – where you want to give more points for greater distances (D) – the points formula for each event is

$$A \times (D - B)^C.$$

You score zero points for a distance equal to or less than B. The distances here are all in metres and the times in seconds. The three numbers A, B, and C are chosen differently for each of the ten events and are shown in Table 3.1. The points achieved for each of the ten events are then added together to give the total score.

In order to get a feel for which events are 'easiest' to score in, take a look at Table 3.2, which shows what you would have to do to score 900 points in each event for an Olympic-winning 9000-point total alongside Ashton Eaton's world record.

There is an interesting pattern in the formulae that change the distances and times achieved into points for the decathlon (Fig. 3.1). The power index C is approximately 1.8 for the running

Table 3.1 Points system variables for the decathlon scoring system

Event	A	B	C
100 m	25.4347	18	1.81
Long jump	0.14354	220	1.4
Shot put	51.39	1.5	1.05
High jump	0.8465	75	1.42
400 m	1.53775	82	1.81
110 m hurdles	5.74352	28.5	1.92
Discus throw	12.91	4	1.1
Pole vault	0.2797	100	1.35
Javelin throw	10.14	7	1.08
1500 m	0.03768	480	1.85

Table 3.2 Event times and distances to score 900 points versus those achieved in Eaton's world record

Event	900 pts	9039 (Eaton, world record)
100 m	10.83 s	10.21 s
Long jump	7.36 m	8.23 m
Shot put	16.79 m	14.20 m
High jump	2.10 m	2.05 m
400 m	48.19 s	46.70 s
110 m hurdles	14.59 s	13.70 s
Discus throw	51.4 m	42.81 m
Pole vault	4.96 m	5.30 m
Javelin throw	70.67 m	58.87 m
1500 m	247.42 s (= 4 min 07.4 s)	254.8 s (= 4 min 14.8 s)

events (1.9 for the hurdles), close to 1.4 for the jumps and pole vault, and close to 1.1 for the throws. The same pattern holds for the women's heptathlon as well, with C approximately equal to 1.8 for runs, 1.75 for jumps, and 1.05 for throws. The fact that $C > 1$ indicates that the points-scoring system is a 'progressive' one, curving upwards in a concave way with decreasing time or increasing distance; that is, the better the performance, the higher the reward for performance improvements (see Fig. 3.1). This is realistic. We know that as you get more expert at your event it gets harder to make the same improvement, but beginners can easily make large gains. The opposite type of points system ('regressive') would have $C < 1$, curving increasingly less, while a 'neutral' one would have $C = 1$ and be a straight line. We can see that the IAAF tables are very progressive for the running events, fairly progressive for the jumps and vault, but almost neutral for the throws.

Figure 3.2 shows the division between the ten events for the averages of the all-time top 100 best-ever men's decathlon performances.

It is clear that there has been a significant bias towards gathering points in the long jump, hurdles, and sprints (100 metres and 400 metres). Performances in these events are all highly correlated with flat-out sprinting speed. Conversely, the 1500 metres and the three throwing events

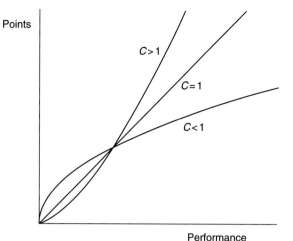

Fig 3.1 Illustration of the increase in points gained for improved performances in progressive ($C > 1$), neutral ($C = 1$), and regressive ($C < 1$) scoring systems.

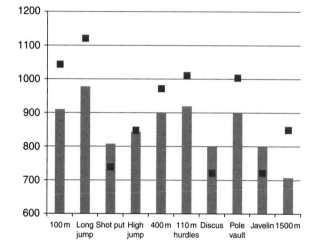

Fig 3.2 Spread of points scored in the ten decathlon events, averaged over the 100 highest-ever points totals. Note that there is significant variance across events. Ashton Eaton's performances in his world record score are marked with boxes. They are unusual: far from the average in every event and they exhibit enormous variances about his mean event score of 903.9 points.

are well behind the other disciplines in points scoring. If you want to coach a successful decathlete, start with a big, strong sprint hurdler and build up strength and technical ability for the throws later. No decathletes bother much with 1500 metres preparation, and rely on general distance-running training. Ideally, we would expect there to be very little variance in this bar chart as we go from one event to another. In reality, the variations are huge and the scoring system is rather biased towards sprinter–jumpers. The women's heptathlon has a similar bias in its scoring tables, which greatly favours sprinter–jumpers like Jessica Ennis over strong throwers.

What if we picked $C = 2$ across all events? This would give an extremely progressive scoring system greatly favouring competitors with outstanding performances in some events (like Eaton), as opposed to those with consistent similar ones. However, it would dramatically favour good throwers over the sprint hurdlers because of the big change from the value of $C = 1.1$ being applied to the throws at present. And this illustrates the basic difficulty with points systems of any sort – there is always a subjective element that could have been chosen differently.

Source

An earlier version of this article appeared on *Maths and sport: Countdown to the games* (<http://sport.maths.org>).

CHAPTER 4

Queen Dido and the mathematics of the extreme

GREG BASON

In 814 BC the Phoenician princess Dido was fleeing across the Mediterranean. Her brother had killed her husband and had denied her a share of the throne of the city-state of Tyre. When she arrived on the coast of North Africa, a local king granted her as much land as could be enclosed by the hide of a single ox. The wise princess ordered the hide to be cut into thin strips, joined together, and arranged to form a circle, for she knew that this shape would give the maximum area of land. This small settlement grew over the years to become the great trading city of Carthage.

So goes the legend. But, as for most foundation myths, there is no evidence for its veracity. Nevertheless, the story does provide one of the earliest recorded examples of a *maximisation* problem; see Fig. 4.1. What is now called *Queen Dido's problem* can be thought of in another way. Imagine all the curves *C* that enclose a given area *A*, and then choose the one that minimises the perimeter. The solution remains a circle but now we have a *minimisation* problem. When mathematicians look to maximise or minimise a quantity, they are solving an *extremal problem*.

The solution to Dido's problem is intuitively a circle. However, it was a very long time after the foundation of Carthage before mathematicians had developed the right tools to *prove* the solution is a circle.

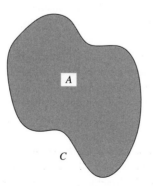

Fig 4.1 Queen Dido's problem: find the maximum area *A* for a given perimeter length *C*.

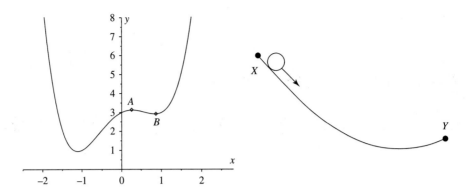

Fig 4.2 (Left) The curve $y = x^4 - 2x^2 + x + 3$, showing a maximum at A and a minimum at B. (Right) The brachistochrone problem: find the curve between X and Y that minimises the time taken by a falling particle.

To understand these developments, we need to move forward 2500 years to 17th-century Europe. Pierre de Fermat was a French lawyer by day and a brilliant mathematician by night. Today he is often regarded as the father of number theory, and his famous *last theorem* remained unsolved for 358 years. Fermat is important to us because he discovered a way of locating the maxima and minima on a curve; see Fig. 4.2 (left). His method *almost* involved the use of derivatives, so Fermat is also remembered for contributing to the early development of *calculus*, as later formalised by Isaac Newton and Gottfried Leibniz (see Chapter 27 for another forgotten hero in the development of calculus). Indeed, Newton famously said, 'If I have seen further it is by standing on the shoulders of giants,' and Newton certainly used the ideas of Fermat when developing his calculus.

Look again at Fig. 4.2 (left). The gradient of the curve changes from point to point. For example, the gradient near point A is close to zero but towards the right-hand side of the curve the gradient is much larger. Maxima and minima occur where the gradient is zero, and to locate these points we need to have a method to calculate the gradient at every point x accurately. The ancients employed mostly *geometrical* methods to solve such extremal problems. However, with the invention of calculus, an alternative and extremely powerful *algebraic* approach became possible. Forming and solving such algebraic equations is now studied by college students all over the world, so we shall not elucidate further. Before proceeding, though, we note that the power of this method should not be understated. If you consider a gradient as something that measures how one quantity changes relative to another, then calculus allows you to model any system that is changing.

It was a generalisation of calculus that enabled mathematicians to finally show that the solution to Queen Dido's problem was indeed a circle. To see why Dido's problem is harder than simply finding the maximum or minimum point of a curve, note that rather than just maximisation over a single variable x, we want to find the maximum area given the endless possibilities for the whole shape of the perimeter; see Fig. 4.1. This is an example from what is known as the *calculus of variations*.

Another famous problem in this field is the *brachistochrone*. Picture a curve lying in a vertical plane and imagine a frictionless bead sliding down the curve from a high point X to a lower one Y; see Fig. 4.2 (right). The brachistochrone problem is to find the curve which minimises the time of descent. The word 'brachistochrone' is from the Greek *brachistos*, meaning 'shortest', and *chronos*, meaning 'time'. Again, to solve this problem we must consider not a single variable but the whole set of possible curves connecting X to Y. The fastest-descent problem was issued as an open challenge to the mathematical community in 1696 by Johann Bernoulli, probably as a way of continuing a bitter feud with his mathematician brother, Jacob. The fastest-descent curve is neither a straight line nor an arc of a circle, and we invite the reader to investigate further.

The brachistochrone problem was solved by several of Bernoulli's contemporaries, but unfortunately their methods did not easily generalise to other problems. However, this changed in 1744 when the Swiss genius, the world's most published scientist, Leonhard Euler, produced a systematic approach to solving calculus-of-variations problems. Some of the ideas used by Euler were formalised by the great Joseph-Louis Lagrange, and in their joint honour the key result in the calculus of variations is named the *Euler–Lagrange equation*.

For the brachistochrone problem we require the function that minimises the descent time, while for Queen Dido's problem we needed the function that maximised the area. Now mathematicians had a way to find both.

So finally, after nearly 3000 years, mathematicians had developed a method for solving Dido's problem, and after some mathematical formalising, Karl Weierstrass provided the first complete proof around 1880. But mathematicians had achieved far more than solving one or two specific problems. They had developed immensely powerful techniques that are used globally every day to solve diverse complex problems. Examples include meteorologists using the calculus of variations to predict the weather and the possible effects of climate change, engineers using it to design structures, physicists considering the interactions of elementary particles, medical scientists trying to understand cancer growth, and economists modelling investment return and market stability.

. . . .

FURTHER READING

[1] Paul Nahin (2007). *When least is best: How mathematicians discovered many clever ways to make things as small (or as large) as possible.* Princeton University Press.
[2] Phil Wilson (2007). Frugal nature: Euler and the calculus of variations. *Plus Magazine*, <http://plus.maths.org/content/frugal-nature-euler-and-calculus-variations>.
[3] Jason Bardi (2007). *The calculus wars: Newton, Leibniz, and the greatest mathematical clash of all time.* Avalon.

Source

This article was a runner-up in a competition run in conjunction with *Mathematics Today*, the magazine of the UK's Institute of Mathematics and its Applications.

CHAPTER 5

Can strings tie things together?

DAVID BERMAN

Unifying forces

To understand the ideas and aims of string theory, it's useful to look back and see how physics has developed from Newton's time to the present day. One crucial idea that has driven physics since Newton's time is that of *unification*: the attempt to explain seemingly different phenomena by a single overarching concept. Perhaps the first example of this came from Newton himself, who in his 1687 work *Principia Mathematica* explained that the motion of the planets in the solar system, the motion of the Moon around the Earth, and the force that holds us to the Earth are all part of the same thing: the force of gravity. We take this for granted today, but before Newton the connection between a falling apple and the orbit of the Moon would have been far from obvious and quite surprising.

The next key unifying discovery was made around 180 years after Newton by James Clerk Maxwell. Maxwell showed that electrostatics and magnetism, which do not appear to be similar at first sight, are just different aspects of a single thing called electromagnetism. In the process Maxwell discovered electromagnetic waves and that these are in fact light – inadvertently explaining yet another seemingly different phenomenon by his unification of electricity and magnetism.

Another two hundred years on, in 1984, Abdus Salam and Steven Weinberg showed that the electromagnetic force and the weak nuclear force, which causes radioactive decay, are both just different aspects of a single force called the electroweak force.

This leaves us with three fundamental forces of nature: gravity, the electroweak force, and the strong nuclear force which holds protons together.

Unifying matter

That deals with the forces, but what about matter? It is an old idea, beginning with the Greeks, that matter is made from a finite number of indivisible elements. This is the original idea of the atom, which modern physics confirms. Experiments performed at CERN in Geneva have shown that there are just twelve basic building blocks of matter. These are known as the *elementary particles*. Everything we've ever seen in any experiment, here or in distant stars, is made of just these twelve elementary particles.

All this is truly impressive: the entire Universe, its matter and dynamics explained by just three forces and twelve elementary objects. It's good, but we'd like to do better. This is where string theory first enters: it is an attempt to unify further. But to understand this, we have to tell another story.

Quantum gravity

There have been two great breakthroughs in 20th-century physics. Perhaps the most famous is Einstein's theory of general relativity. The other equally impressive theory is quantum mechanics.

General relativity is itself a unification. Einstein realised that space and time are just different aspects of a single object he called *spacetime*. Massive bodies like planets can warp and distort spacetime, and gravity, which we experience as an attractive force, is in fact a consequence of this warping. Just as a pool ball placed on a trampoline will create a dip that a nearby marble will roll into, so does a massive body like a planet distort space, causing nearby objects to be attracted to it.

The predictions made by general relativity are remarkably accurate. In fact, most of us will have inadvertently taken part in an experiment that tests general relativity: if it were false, then global positioning systems would be wrong by about 50 metres per day. The fact that GPS works to within five metres in ten years illustrates the accuracy of general relativity.

The other great breakthrough of the 20th century was quantum mechanics. One of the key ideas here is that the smaller the scale at which you look at the world, the more random things become. Heisenberg's uncertainty principle is perhaps the most famous example of this. The principle states that when you consider a moving particle, for example an electron orbiting the nucleus of an atom, you can never ever measure both its position and its momentum as accurately as you like. Looking at space at a minuscule scale may allow you to measure position with a lot of accuracy, but there won't be much you can say about momentum. This isn't because your measuring instruments are imprecise. There simply isn't a 'true' value of momentum, but a whole range of values that the momentum can take, each with a certain probability. In short, there is randomness. This randomness appears when we look at particles at a small enough scale. The smaller one looks, the more random things become!

The idea that randomness is part of the very fabric of nature was revolutionary: it had previously been taken for granted that the laws of physics didn't depend on the size of things. But in quantum mechanics they do; the scale of things does matter. The smaller the scale at which you look at nature, the more different from our everyday view of the world it becomes: and randomness dominates the small-scale world.

Again, this theory has performed very well in experiments. Technological gadgets that have emerged from quantum theory include the laser and the microchip that populate every computer and mobile phone.

But what happens if we combine quantum mechanics and relativity? According to relativity, spacetime is something that can stretch and bend. Quantum mechanics says that on small scales things get random. Putting these two ideas together implies that on very small scales spacetime itself becomes random, pulling and stretching, until it eventually pulls itself apart.

Evidently, since spacetime is here and this hasn't happened, there must be something wrong with combining relativity and quantum mechanics. But what? Both of these theories are well tested and believed to be true. Perhaps we have made a hidden assumption?

It turns out that indeed we have. The assumption is that it's possible to consider smaller and smaller distances and get to the point where spacetime pulls itself apart. What has rested in the back of our minds is that the basic indivisible building blocks of nature are point-like, but this may not necessarily be true.

Strings to the rescue

This is where string theory comes to the rescue. It suggests that there is a smallest scale at which we can look at the world: we can go that small but no smaller. String theory asserts that the fundamental building blocks of nature are not like points, but like strings: they have extension; in other words, they have length. And that length dictates the smallest scale at which we can see the world.

What possible advantage could this have? The answer is that strings can vibrate. In fact, they can vibrate in an infinite number of different ways. This is a natural idea in music. We don't think that every single sound in a piece of music is produced by a different instrument; we know that a rich and varied set of sounds can be produced by even just a single violin. String theory is based on the same idea. The different particles and forces are just the fundamental strings vibrating in a multitude of different ways.

The mathematics behind string theory is long and complicated, but it has been worked out in detail. But has anyone ever seen such strings? The honest answer is 'no'. The current estimate of the size of these strings is about 10^{-34} metres, far smaller than we can see today, even at CERN. Still, string theory is so far the only known way to combine gravity and quantum mechanics, and its mathematical elegance is for many scientists sufficient reason to keep pursuing it.

The theory's predictions

If string theory is indeed an accurate model of spacetime, then what else does it tell us about the world?

One of its more startling and most significant predictions is that spacetime is not 4-, but 10-dimensional. It is only in 10 dimensions of spacetime that string theory works. So where are those six extra dimensions? The idea of hidden dimensions was in fact put forward many years before the advent of string theory by the German mathematician Theodor Kaluza and the Swedish physicist Oskar Klein.

Shortly after Einstein described the bending of space in general relativity, Kaluza and Klein considered what would happen if a spatial dimension were to bend round and rejoin itself to form a circle. The size of that circle could be very small, perhaps so small that it couldn't be seen. Those dimensions could then be hidden from view. Kaluza and Klein showed that in spite of this, these dimensions could still have an effect on the world we perceive. Electromagnetism becomes a consequence of the hidden circle, with motion in the hidden dimension being electric charge.

Hidden dimensions are possible, and they can in fact give rise to forces in the dimensions that we can see.

String theory has embraced the Kaluza–Klein idea and currently various experiments are being devised to try and observe the hidden dimensions. One hope is that the extra dimensions may have left an imprint on the cosmic microwave background, the left-over radiation from the Big Bang, and that a detailed study of this radiation may reveal them. Other experiments are more direct. The force of gravity depends crucially on the number of dimensions, so by studying gravitational forces at short distances one can hope to detect deviations from Newton's law and again see the presence of extra dimensions.

Mathematics and physics have always influenced each other, with new mathematics being invented to describe nature and old mathematics turning out to lend perfect descriptions for newly discovered physical phenomena. String theory is no different, and many mathematicians work on ideas inspired by it. These include the possible geometries of the hidden dimensions, the basic ideas of geometry when there is a minimum distance, the ways in which strings can split and come together, and the question of how we can relate strings to the particles in the world that we see.

String theory gives us an exciting vision of nature as minuscule bits of vibrating string in a space with hidden curled-up dimensions. All the implications of these ideas are yet to be understood. String theory is an active area of research, with hundreds of people working to see how the theory fits together and produces the world we see around us. It will be very exciting to see how it develops in the next fifty years!

Source

An earlier version of this article appeared in *Plus Magazine* (<http://plus.maths.org>) on 1 December 2007.

CHAPTER 6

Dimples, grooves, and knuckleballs

KEN BRAY

What makes an ideal football? The ball must be perfectly round and retain its shape and internal pressure after a lot of physical abuse. It should be bouncy, but not too lively when kicked or headed and it must not absorb water. And, finally, it should move about in a pacy manner when passed between the players and be capable of impressive turns of speed for shots at goal.

The last point may seem trivial: surely footballs go faster the harder they're kicked? This is broadly true, but how much of this initial speed is retained in flight depends critically on what happens as air flows around the ball's surface. More research is devoted to this factor than any other aspect of the ball's design.

Flow separation and the drag crisis

The images in Fig. 6.1 show a model ball mounted in a wind tunnel. The ball is stationary, but by varying the speed of the airflow around it we can mimic what happens when the ball is in flight. Flow patterns are made visible by introducing a little smoke into the airstream.

At very low speeds the flow follows the surface of the ball intimately, but as the speed is increased, as in the image on the left of Fig. 6.1, the flow begins to break away at the aptly named *separation points* (indicated by the arrows). Note that when separation occurs at lowish speeds

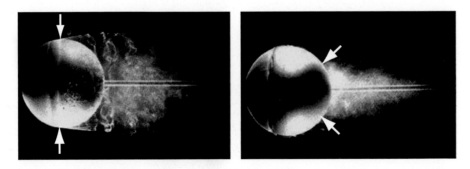

Fig 6.1 Flow separation and transition points for a model ball in a wind tunnel. The image on the left shows early separation and high drag, while that on the right shows late separation and low drag.

Images by Henri Werlé, copyright Onera (<http://www.onera.fr/en>), the French Aerospace Lab.

there is a very large turbulent region behind the ball, clearly visible in the image. Turbulence causes aerodynamic drag, effectively robbing the ball of kinetic energy.

However, when the speed is increased a surprising thing happens. The separation points move towards the rear of the ball (as in the image on the right of Fig. 6.1) and the flow clings to the surface once more. Downstream turbulence is greatly reduced, implying a significant reduction in aerodynamic drag. This transition is very sudden and is enormously important in ball games. It is caused by surface roughness: it seems very counter-intuitive that a slightly rough surface should have a superior aerodynamic performance compared with a smooth one.

Aerodynamic drag can be measured very precisely by using sensitive instruments attached to the rig on which the ball is mounted. The drag force F is then characterised using a simple expression

$$F = \frac{1}{2} C_\mathrm{d} \rho A V^2.$$

Here ρ is the density of air, A the cross-sectional area of the ball, and V its speed (or the speed of the airflow in the wind tunnel). The parameter C_d is the *drag coefficient*, a number that scales the strength of the drag force at a given speed.

Knowing the variation of C_d with speed is very important in ball design, and determining its value is quite simple: F is measured at various speeds and then C_d is calculated from the above expression. The graph in Fig. 6.2 shows some experimental results obtained in this way for the classic 32-panel ball (a surface pattern of 12 pentagons and 20 hexagons) (by researchers in Japan) and for a smooth ball (by researchers in Germany).

The switch from high to low C_d at some critical speed is very sudden, and for this reason the transition is often referred to as the *drag crisis*. Notice also in this graph that the switch for the 32-panel ball occurs at a much lower speed than that for a smooth ball, about 12 metres per second (27 miles per hour) compared with 37 metres per second (83 miles per hour). This is very fortunate, as most of the important actions in football such as long throw-ins, corners, and free kicks take place within the low-drag regime after this switch occurs.

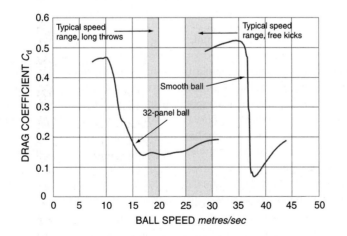

Fig 6.2 Variations in the drag coefficient C_d with different ball speeds. The typical speed ranges for long throws and free kicks are shown in the shaded regions.

As already noted, surface roughness is responsible for this important transition, and it was not realised for many years that the stitched seams between the ball's panels produced just the right degree of aerodynamic roughness to do the trick. In fact, it would be almost physically impossible to play a decent game of football with a perfectly smooth ball, since only in very exceptional circumstances would it move fast enough to avert the drag crisis.

Ancient maths for modern footballs

So seams and panels are important. It would be useful to know exactly how much seam there is on a modern ball and also whether any other configuration of ball panels can give a better result. On a classic 32-panel ball, each side of each pentagon or hexagon is shared with its neighbour. So if we calculate the total perimeter of all the hexagons and pentagons and divide by two we'll have our answer. There are 20 hexagons and 12 pentagons, so the combined perimeters add up to $6 \times 20 + 5 \times 12 = 180$ sides and dividing by two gives us 90 individual portions of seam. The total length of seams on the ball, L, is just

$$L = 90s,$$

where s is the length of each side of a hexagon or pentagon. At this point, we might simply measure s using a classic 32-panel ball inflated to the correct pressure: the average of careful measurements of 10 portions of seam gives us a length of 4.57 centimetres. So we find the total length of seam is $L = 90s = 4.12$ metres.

Rather than measuring a real ball, we could also calculate an approximation for s. Taking the radius of our ball to be R (which is 11 centimetres for a decent football), we can approximate the surface area of our ball by the patchwork of pentagons (with area $A_p = \frac{\sqrt{25+10\sqrt{5}}}{4}s^2$) and hexagons (with area $A_h = \frac{3\sqrt{3}}{2}s^2$):

$$20A_h + 12A_p = 4\pi R^2.$$

This gives us a simple equation for s and, solving it, we find $s = 4.58$ centimetres, a close approximation to the length we measured for a real ball.

Is there another configuration of panels that would give us more seam? The answer comes from a famous expression first obtained by the Swiss mathematician Leonhard Euler. Ignoring its surface curvature, the 32-panel ball corresponds to a *truncated icosahedron*, one of the 13 classic polyhedra known as the *Archimedean solids* (Fig. 6.3). The faces of these are all regular polygons such as triangles, squares, pentagons, and hexagons. For these objects Euler's formula states that $V + F - E = 2$, where E and F are the numbers of edges and faces and V is the number of corners, called *vertices*, of the solid.

The formula tells us that for greater seam length, the monster known as the *snub dodecahedron* could be used. Counting the faces (80 triangles, 12 pentagons) and the vertices (60), we find from Euler's formula that the number of edges is 150. Then, using a similar calculation to that for the 32-panel ball described above, it turns out that the snub dodecahedron soccer ball would have a total seam length of very nearly 9.5 metres. Of course, no manufacturer would make such a ball:

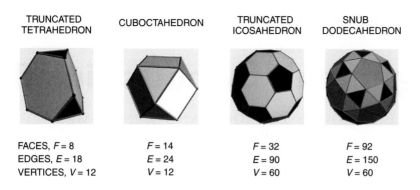

TRUNCATED TETRAHEDRON	CUBOCTAHEDRON	TRUNCATED ICOSAHEDRON	SNUB DODECAHEDRON
FACES, $F = 8$	$F = 14$	$F = 32$	$F = 92$
EDGES, $E = 18$	$E = 24$	$E = 90$	$E = 150$
VERTICES, $V = 12$	$V = 12$	$V = 60$	$V = 60$

Fig 6.3 Selection of Archimedean solids. Images created by Robert Webb's Great Stella software.

joining 92 panels accurately would be far too fiddly a process and, in any case, the 32-panel ball performs perfectly well.

Grip and groove or slip and move?

If we go the other way and reduce the number of panels, we have the problem of less implicit surface roughening. A smoother surface does not appear to be the cleverest design objective for a modern ball as the low-drag regime doesn't kick in until the ball is moving at very high speeds, far higher than the ball speeds in a typical game. And yet, since 2006 this has been the driving force, with panel numbers tumbling from 32 to 14 for the 2006 World Cup and then to only eight for the World Cup held in South Africa in 2010.

At the 2006 World Cup goalkeepers complained bitterly about the unpredictable movement of the official ball, called the Teamgeist (German for 'team spirit'). When the ball was kicked fast with very little spin, as the individual portions of the seam rotated into the airflow, the flow separation points (such as those shown in the wind tunnel images in Fig. 6.1) switched positions and the ball bobbed about erratically. This movement came to be known as 'knuckling', a baseball term referring to a pitcher's attempt to throw the ball with scarcely any spin. The very limited seam structure of a baseball (two panels, one seam) practically guarantees aerodynamic instability and the knuckleball's unpredictable movement is very testing for batters in baseball.

Artificial roughening was used to fix the problem, as seen in the Jabulani (Zulu for 'to celebrate') used in the 2010 World Cup. This ball had only eight panels, so surface texturing was a must and the panels were covered with an intricate pattern of very fine surface depressions. The Jabulani appears to behave more stably in flight than the Teamgeist, but two things seem certain. For major tournaments, there will be no return to the multiple-panel formats of the past; in fact, we might expect even fewer panels than the eight of the present day. And, necessarily, careful surface texturing will be needed to compensate for the seam roughening lost as a consequence.

Source
An earlier version of this article appeared as 'A fly walks round a football' in *Plus Magazine* (<http://plus.maths.org>) on 15 September 2011.

CHAPTER 7

Pigs didn't fly but swine flu

ELLEN BROOKS-POLLOCK AND KEN EAMES

The little town of La Gloria, nestled in the hills to the east of Mexico City, is a fairly unremarkable place. With a population of a little over 2000, many of whom commute to the city during the week, and a large pig farm a few miles away, it had little to bring it to the world's attention. Until, that is, March 2009, when a new strain of flu was identified in the Americas. The new virus contained elements of human, avian, and swine influenzas, and appeared to have originated in Mexico. The nearby pig farm, the links to Mexico City, and the fact that about 600 residents reported respiratory symptoms led to La Gloria being suggested as the place where it all began.

Whether or not swine flu really started in La Gloria, it quickly became a global problem. Mexico was initially hardest hit and from there the new virus began to spread. By the end of April it was in the UK, carried by returning holidaymakers, and before long it had reached every part of the globe. Health authorities around the world swung into action: public health campaigns were launched, schools closed, supplies of antiviral drugs made ready, and vaccines ordered.

About a million people in England were infected; one of us (Ellen) managed to avoid it, but the other (Ken) wasn't so lucky. He doesn't know who he got it from, but he knows that he passed it on to his partner, who made the mistake of bringing him tea and sympathy as he coughed and spluttered in bed for a couple of days. For mathematical modellers who spend most of their time creating simulations of epidemics, this practical experience of swine flu in action provided a useful, if unwelcome, dose of reality.

Epidemic modelling

In 1927 William Kermack and Anderson McKendrick, two scientists working in Edinburgh, published a paper describing the now well-known *Susceptible–Infected–Recovered* (SIR) model for describing how a new disease spreads in a population. In searching for a mechanism to explain when and why an epidemic terminates, they made the key observation that the progress of epidemics rests on a single number from their model: 'In general a threshold density of population is found to exist, which depends upon the infectivity, recovery and death rates peculiar to the epidemic. No epidemic can occur if the population density is below this threshold value.'

The realisation that the number of *unaffected* (that is, susceptible), rather than infectious, people governed epidemic dynamics had profound consequences for controlling disease outbreaks. For instance, vaccination strategies depend on the minimum number of susceptible people required for an epidemic; exceed this threshold and there is the potential for an outbreak.

The insights of Kermack and McKendrick were the first steps towards understanding disease dynamics, but the real world is much more complex and interesting. The 2009 swine flu epidemic demonstrated the importance of population structure, in addition to population size, for describing the progression of a disease through a population.

When describing an epidemic mathematically, most modellers will start with a set of equations similar to those of Kermack and McKendrick. In a population of size N, people are assumed to be either susceptible, infectious, or recovered (and immune), with $S(t), I(t)$, and $R(t)$ giving the number of people in each of these groups at a time t.

The simplest models have two parameters: the *recovery rate*, γ, and the *transmission rate*, β. In the early stages of an outbreak, much effort goes into estimating these rates. The recovery rate, γ, can be measured by closely monitoring the first few cases. The transmission rate, β, is hard to measure directly as it captures both the physical process of meeting someone and the biological process of transmitting infection; in the case of flu, measuring β is even more challenging since many infections are mild and go undetected.

The basic model assumes that everyone in the population has constant and equal contact with everyone else. Therefore, each infectious person causes new infections at a rate $\beta S(t)/N$, where the $S(t)/N$ term accounts for the fact that not all the people that an infectious person meets will be susceptible. In order for an epidemic to grow, the number of new infections must exceed the number of people recovering from the disease; that is, $\beta I(t)S(t)/N > \gamma I(t)$. This important threshold demonstrated by Kermack and McKendrick as necessary for an epidemic is described in terms of the *basic reproductive number* of a disease, $R_0 := \beta/\gamma$. The number R_0 represents the average number of new infections produced by an average infectious person in an otherwise susceptible population. From an individual's point of view, the threshold is equivalent to saying that on average each infectious person must produce more than one secondary case, or, at the start of an outbreak, that $R_0 > 1$.

Modelling swine flu in the UK

From the early stages of the swine flu epidemic, the basic reproductive number R_0 was estimated as approximately 1.3 and the recovery rate γ as approximately 0.5 per day. Taking the initial

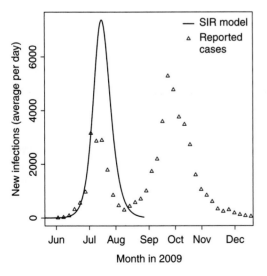

Fig 7.1 The epidemic curve produced by the standard SIR model with swine-flu-like parameters ($\gamma = 0.5$ per day, $R_0 = 1.3$, reporting rate = 0.1, and $N = 60$ million), solid line. The actual number of reported cases is shown for comparison.

number of infectious individuals as $I(t = 0) = 60$, i.e. initially there were around 60 infected people out of a total population of 60 million, Fig. 7.1 compares the model (scaled by the fraction of cases that sought medical attention, estimated at 0.1) with the number of reported cases in England.

The epidemic produced by the standard SIR model clearly does not reproduce the observed epidemic. Most notably, the model predicts a single peak (late July), whereas the real epidemic curve has two peaks (mid July and late September). To reproduce the dynamics of this epidemic we must revisit how the transmission of the infection was modelled.

The importance of social mixing for explaining the two epidemic waves

One way to generate multiple epidemic peaks is for the transmission rate, β, to vary over time. As mentioned above, β encompasses multiple processes. If we assume that the *biological* processes of transmission and recovery are fixed, then it must be the way that people interact – the social mixing – that is changing.

The peak of the summer epidemic wave occurred in the week in which most schools in England broke up for their summer holidays, suggesting that the decline in the epidemic was caused by altered contact rates, resulting in a change in β. Essentially, the basic reproductive number was lower during school holidays than during term time; in particular, during the school holidays each infected person caused fewer than one secondary case on average, so the epidemic decreased.

A further consequence of the contact patterns seen in the real world arises in the form of different rates of infection in adults and children. From surveys, we know that school-aged children have almost twice as many social interactions as adults, indicating that infectious children typically generate more new infections than adults. Further, the vast majority of social contacts are between people of a similar age, so infection is concentrated in children.

A representation of how the population is connected socially is shown in Fig. 7.2. Each line represents a social contact that is a potential transmission route for the infection. Children (circles) are significantly more strongly connected to other children than adults (squares) are to other adults, and children and adults are only loosely connected to each other.

We can improve the realism of the standard SIR model by splitting the population into children and adults and creating two linked submodels: one for the progression of the infection through children and one for adults. Each of these submodels is dependent on the mixing within and between the subpopulations. We allow the contact rates to fall during the summer holidays (indicated by the shaded region in Fig. 7.3). Here, we have used information about social contacts collected from Flusurvey, an online surveillance system set up during the pandemic (<http://www.flusurvey.org.uk>). Apart from these changes to the transmission rate, the parameters remain the same as before: $\gamma = 0.5$ per day, the term-time reproductive number R_0 is approximately 1.3, and the fraction of cases that seek medical attention is 0.1 in a population of 20 million children and 40 million adults. For further details, see [2].

The output of this model is shown in Fig. 7.3. This new model reproduces two main features of the epidemic: the two peaks caused by changing contact patterns over the summer and the lower epidemic in adults resulting from their lower levels of social interactions.

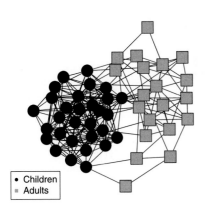

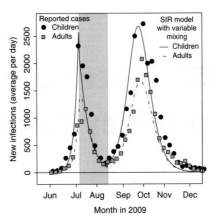

Fig 7.2 An age-structured population in which children (circles) have more connections than adults (squares).

Fig 7.3 The epidemic curve produced by the age-structured SIR model, incorporating a fall in social contacts during the holidays.

The next stages in epidemic modelling

Mathematical epidemic modelling has come a long way since Kermack and McKendrick's groundbreaking work in the 1920s. Modern computing power has allowed researchers to develop massively complex models, containing billions of unique, interacting individuals, and the sorts of models routinely used these days are far more sophisticated than there is space here to describe properly. (See also Chapter 19 for mathematical modelling of virus replication and mutation at the molecular level.) Nevertheless, as we have seen, even simple models can go a long way towards explaining observed epidemic patterns. Simple or complex, these models will be governed by the answers to a few key questions: How long does an infection last? How quickly does it spread? How does it get from person to person? Good data – from surveys, from surveillance, and from field epidemiology – are needed to answer these questions.

. . . .

FURTHER READING

[1] William Kermack and Anderson McKendrick (1927). Contribution to the mathematical theory of epidemics – I. *Proceedings of the Royal Society A*, vol. 115, pp. 700–721.
[2] Ellen Brooks-Pollock and Ken Eames (2011). Pigs didn't fly, but swine flu. *Mathematics Today*, vol. 47, pp. 36–40.

Source

A longer version of this article appeared in *Mathematics Today* in February 2011.

CHAPTER 8

Bill Tutte: Unsung Bletchley hero

CHRIS BUDD

Alan Turing, whose centenary was celebrated in 2012, is rightly applauded as the man who both played a major role in cracking the German Enigma code during the Second World War and also as being one of the fathers of modern electronic computing (see also Chapter 48 for his less well-known contribution to biology). However, he was not the only code breaker working at the secret establishment in Bletchley Park. Indeed, there were at least 12,000 others, many of whom were great mathematicians in their own right (including Professor David Rees FRS, one of the founders of the IMA). Amongst the greatest of these was the mathematician Bill Tutte OC, FRS (14 May 1917–2 May 2002), whose extraordinary feat in breaking the fiendishly hard Tunny code has been described as one of the greatest intellectual achievements of the war. His portrait is in Fig. 8.1.

Bill Tutte started his mathematical career (as did many others, including Sir Isaac Newton and Bertrand Russell) at Trinity College, Cambridge. Whilst he was an undergraduate, he was one of a team of students who were collectively known as 'Descartes' and who tackled a lovely problem known as *squaring the square*. This involves finding a set of squares with integer sides which together tile a larger square. Try it for yourself: there are boring solutions such as using unit squares, and more interesting variants such as tilings with squares that have different integer sides. Like many of the best maths problems, it is easy to state but hard to solve. The Trinity team found a solution using ideas from electronic circuit theory (showing that even pure mathematics can learn

Fig 8.1 Bill Tutte (1917–2002). Copyright *Newmarket Journal*.

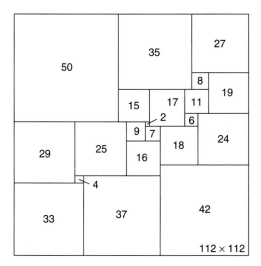

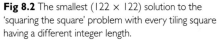

Fig 8.2 The smallest (122 × 122) solution to the 'squaring the square' problem with every tiling square having a different integer length.

something from engineering). Fittingly, the solution of the problem using the smallest possible number of different-sided squares is now the logo of the Trinity Maths Society; see Fig. 8.2.

Like many Cambridge mathematicians, Tutte joined the secret code-breaking establishment, the Government Code and Cipher School at Bletchley Park, near Milton Keynes, at the start of the war. He worked in the Research Section, led by John Tiltman. Whilst most of Bletchley was heavily engaged in breaking the Enigma codes which were used to encode the day-to-day messages of the German armed forces, the research group was seeking to break some of the other German ciphers, some of which were being used by Hitler himself. One of these was the cipher produced by the German Lorenz SZ 40/42 encryption machine, based on the then modern teleprinter machines and using a Vernan stream cipher, the principle being that a series of 12 rotors were attached to a teleprinter machine which combined its output with a pseudo-random secret key generated by the movements of the rotors. This was a far more sophisticated machine than the (relatively simple) rotor-based encipherment machines used in the Enigma codes.

At the time it seemed impossible that the code breakers at Bletchley would ever break this cipher, code-named Tunny. Indeed, unlike the situation for Enigma, they had never even seen a Lorenz machine and knew none of its internal workings. However, the potential rewards were very great, as if they could break it then they could read Hitler's most secret orders themselves.

A breakthrough came on 30 August 1941 when, following a mistake by an operator, the Germans sent almost the same message twice on the same machine and with the same key setting, a practice that was strictly forbidden. This was picked up by the secret British 'Y' receiving station in Kent and sent to Bletchley, where it was studied by Tiltman. Knowing that the two messages were almost identical, Tiltman was able to decipher (by a process of subtraction and careful comparison) the two enciphered messages, but, even more importantly, he was able to find 4000 characters of the secret key.

In a sense, so what? One message had been decoded, but what about all of the others? The job of the Research Section was to find out how the Lorenz machine worked, and in particular to work out how the pseudo-random key was generated. If they could do this, then in principle they could work out how all of the pseudo-random keys were generated, and would then be able to decode messages in the future.

The job of doing this was given to Tutte, who proceeded to write out the original teleprinter code and the key by hand and then to look for possible repetitions and regularities in it caused by the movements of the code wheels. Splendidly, he did not use a machine to do this, but instead he wrote out the key on squared paper with a new row after a defined number of characters that was suspected of being the frequency of repetition of the key. If this number was correct, the columns of the matrix would show more repetitions of sequences of characters than chance alone. Following a series of hunches, and observations, he tried with a period of 41 and 'got a rectangle of dots and crosses that was replete with repetitions'. This showed him that the most rapidly moving wheel in the key-generating sequence had 41 teeth. So was 41 the answer? Not quite; in fact, Tutte had to work out the sequences of impulses of all of the rotors, not all of which moved in the same way. He deduced that there were two types of rotor, which he called χ and ψ, each of which moved in different ways, and which combined together to generate the key.

Following Tutte's breakthrough the rest of the Research Section at Bletchley joined in the programme of cracking the key, and together they found out that the five ψ wheels all moved together under the control of two more wheels, called μ. At the end of this intense period of work, the complete inner workings of the Lorenz machine had been deduced without such a machine ever being observed directly.

What makes this remarkable calculation even more extraordinary is that Tutte was only 24 at the time!

After the key-generating mechanism had been cracked, an enormous amount of work still had to be done to crack Tunny. A special team of code breakers was set up under Ralph Tester, which became known as the Testery. This team included both Tutte and Donald Michie (who was of a similar age to Tutte and who went on to become one of the key founders of modern computing and artificial intelligence). Turing also made a significant impact on the Testery's work.

This team performed the bulk of the subsequent work in breaking Tunny messages using hand-based methods, again largely based on Tutte's ideas, which exploited statistical regularities in the key caused by combined movements of some of the rotors. However, the sheer difficulty of cracking Tunny made it clear that further, automatic methods were needed.

Therefore a complementary section under Max Newman, known as the Newmanry, was set up which developed code-breaking machines. The first of these were called Robinsons (after Heath Robinson), but the huge achievement of the Newmanry was the development of electronic-valve-based code-breaking computers. The most important of these was the Colossus computer, designed by Tommy Flowers. The final version of this machine had 2400 valves and processed messages at a speed of 25,000 characters per second, allowing the most secret German messages to be decoded just in time for D-Day. A rebuilt (and operational) Colossus machine is now on view at Bletchley Park.

After the war Tutte went back to Cambridge to complete his PhD under the supervision of another Bletchley Park veteran called Shaun Wylie (who, coincidentally, also taught me when I was at Cambridge, though much later). Tutte then moved to Canada, where he got married and stayed for the rest of his life, first in Toronto and then in Waterloo, where he founded the Department of Combinatorics and Optimization. His mathematical career concentrated on combinatorics, the study of how objects are linked together, and this work was possibly inspired not only by his code-breaking activities at Bletchley but also his earlier work on squaring the square. A hugely important part of combinatorics is graph theory, which he is credited as having helped create in its modern form, and which allows us to study networks. This is arguably one of the most important applications of mathematics to 21st-century technology, as the applications of network theory include the Internet, Google, social networks, and mobile phone systems, as well as its application to the four colour problem (and the Königsberg bridges problem, see Chapter 20). Tutte also worked in the deep mathematical field of matroid theory, proving many important results. For this and related work he was elected a Fellow of the Royal Society in 1987, and to the Order of Canada in 2001. Because of his seminal work in breaking Tunny, it was a fitting tribute to him that in 2011 Canada's Communications Security Establishment named an organisation aimed at promoting research into cryptology the Tutte Institute for Mathematics and Computing. He and other mathematicians like him not only helped to win the war, but have made the world a better place in many other ways.

CHAPTER 9

What's the use of a quadratic equation?

CHRIS BUDD AND CHRIS SANGWIN

In 2003 the quadratic equation was pilloried at a UK National Union of Teachers conference as an example of the cruel torture inflicted by mathematicians on poor, unsuspecting school children. The proclamation was that it and mathematics were useless; no one wanted to study mathematics anyway, so why bother? So is the quadratic equation really dead? Perhaps, but it really isn't the quadratic equation's fault. In fact, it has played a pivotal part in the whole of human civilisation, saving lives along the way.

It all started around 3000 BC with the Babylonians. Amongst their many inventions was the (dreaded) taxman, giving a reason why they needed to solve quadratic equations. Imagine that a Babylonian farmer had a square crop field (not that a square gives the maximal area for a given perimeter length; see Chapter 4). What amount of crops could be grown there? Double the side length of the field, and *four* times as much of the crop could be grown than before. The reason for this is that the crop quantity potentially grown is proportional to the *area* of the field, which in turn is proportional to the *square* of the side length. Mathematically, if x is the side length of the field, a is the amount of crop that could be grown on a square field of unit side length, and c is the amount that can be grown in our particular field, then

$$c = ax^2.$$

The taxman arrives; cheerily he says to the farmer, 'I want you to give me c crops to pay for the taxes on your farm.' The farmer now has a problem: how big a field needs growing for that crop amount? We can answer this question easily; in fact,

$$x = \sqrt{\frac{c}{a}}.$$

Fig 9.1 (Left) A non-rectangular Babylonian field. (Right) A rectangle with side lengths proportional to the golden section.

Now, not all fields are square. Suppose that the farmer had a more oddly shaped field with two triangular sections – see Fig. 9.1. For appropriate values of a and b, the amount of crop the farmer could grow is given by

$$c = ax^2 + bx.$$

This looks a lot more like the usual quadratic equation, and even under the taxman's eye, it's a lot harder to solve. Yet the Babylonians came up with the answer again. First, we divide by a and add $b^2/4a^2$ to both sides to obtain

$$\left(x + \frac{b}{2a}\right)^2 = \frac{c}{a} + \frac{b^2}{4a^2}.$$

This equation can be solved by taking square roots, the result being the famous '$-b$ formula',

$$x = -\frac{b}{2a} \pm \sqrt{\frac{c}{a} + \frac{b^2}{4a^2}} \quad \text{or} \quad x = \frac{-b \pm \sqrt{b^2 + 4ac}}{2a}.$$

The fact that taking a square root can give a positive or a negative answer leads to the remarkable result that a quadratic equation has *two* solutions.

Now, this is where the teaching of quadratic equations often stops. We have reached the journalists' beloved object when they interview mathematicians – a *formula*. Endless questions can be devised which involve different values of a, b, and c that give (two) answers. However, this isn't what mathematics is about at all. Finding a formula is all very well, but what does the formula *mean*; does having a formula really matter?

We fast-forward 1000 years to the ancient Greeks. They were superb mathematicians and discovered much that we still use today. One of the equations they wished to solve was the (simple) quadratic equation $x^2 = 2$. They knew that this equation had a solution. In fact, it is the length of the hypotenuse of a right-angled triangle with equal shorter sides, both of length 1.

So, what is x here? Or, to ask the question that the Greeks asked, what *sort* of number is it? Why this mattered lay in the Greeks' sense of *proportion*. They believed that all numbers were in proportion with each other, or, all numbers could be written as ratios of two whole numbers. Hence, surely it was natural that $\sqrt{2}$ was such a ratio. However, it isn't. In fact,

$$\sqrt{2} = 1.4142135623730950488\ldots,$$

and the decimal expansion of $\sqrt{2}$ continues without any discernible pattern; $\sqrt{2}$ was the first *irrational number* (a number impossible to write as a ratio) to be recognised as such. Other examples include $\sqrt{3}$, π, e, and in fact 'most' numbers. This discovery caused both great excitement and shock, with the discoverer committing suicide. (A stark warning to the mathematically keen!) At this point, the Greeks gave up algebra and turned to geometry.

We in fact meet $\sqrt{2}$ regularly. In Europe paper sizes are measured in A sizes, with A0 being the largest, with an area of precisely 1 m^2. The A sizes have a special relationship between them. If we take a sheet of A0 paper and then fold it in half (along its longest side), we obtain A1 paper. Folding it in half again gives A2 paper, etc. However, the paper is designed such that the proportions of each of the A sizes are the same; each piece of paper has the same shape.

What proportion is this? Start with a piece of paper with sides x and y, with x being the longest side. Now divide it into two to give another piece of paper, with sides y and $x/2$, with y now being the longest side. The proportions of the first piece of paper are x/y and those of the second are $y/(x/2)$ or $2y/x$. We want these two proportions to be equal. Hence,

$$\frac{x}{y} = \frac{2y}{x} \quad \text{or} \quad \left(\frac{x}{y}\right)^2 = 2.$$

Another quadratic equation! Fortunately it's one we have already met. Solving it, we find that

$$\frac{x}{y} = \sqrt{2}.$$

This result is easy to check. Just take a sheet of A4 paper and measure the sides. We can also work out the size of each sheet. The *area A* of a piece of A0 paper is given by

$$A = xy = x\left(\frac{x}{\sqrt{2}}\right) = \frac{x^2}{\sqrt{2}}.$$

However, we know that $A = 1$ m^2, so we have another quadratic equation for the longest side x of A, given by

$$x^2 = \sqrt{2} \text{ m}^2 \quad \text{or} \quad x = \sqrt{\sqrt{2}} \text{ m} = 1.189207115\ldots \text{ m}.$$

This means that the longest side of A2 is given by $x/2 = 59.46$ cm and that of A4 by $x/4 = 29.7$ cm. Check these on your own sheets of paper.

Paper used in the United States, with dimensions of 8.5 inches by 13.5 inches, and called *foolscap*, has a different proportion. To see why, we return to the Greeks. Having caused such grief, the quadratic equation returns in the search for the perfect proportions: a search that continues today in the design of film sets.

Let's start with a rectangle. We remove a square from it with the same side length as the shortest side of the rectangle (see the right-hand plot in Fig. 9.1). If the longer and shorter sides have lengths 1 and x respectively, the square has sides of length x. Removing it gives a smaller rectangle with longer side x and smaller side $1 - x$. So far, so abstract. However, the Greeks believed that the rectangle which had the most aesthetic proportions was that for which the large and the small rectangles constructed above had the same proportions. For this to be possible we must have

$$\frac{x}{1} = \frac{1-x}{x} \quad \text{or} \quad x^2 + x = 1.$$

This is yet another quadratic equation; it has the (positive) solution

$$x = \frac{\sqrt{5} - 1}{2} = 0.61803\ldots$$

The number x is called the *golden ratio* and is often denoted by the Greek letter ϕ (pronounced 'phi'). It is certainly true, and immensely interesting, that it is the most irrational number possible (that means it is very hard to approximate it by a fraction). It is also rumoured to be important in the understanding of beauty, although, as in many things, the truth of this may be mainly in the eye of the beholder.

Before his infamous run-in with the Inquisition, Galileo Galilei devoted much of his life to studying motion, in particular the understanding of dynamics. This has huge relevance to such vital everyday activities as knowing when to stop your car. Central to this is the idea of *acceleration*, which leads directly to quadratic equations.

If an object is moving in one direction without a force acting on it, it continues to move in that direction with a constant velocity. Let's call this velocity v. Now, if the object starts at the point $x = 0$ and moves like this for a time t, the resulting position is given by $x = vt$. Usually the object has a force acting on it, such as gravity for a projectile or friction when a car brakes. Fast-forwarding to Newton, we know that the effect of a constant force is to produce a constant acceleration a. If the starting velocity is u, then the velocity v after a time t is given by $v = u+at$. Galileo realised that from this expression he could calculate the position of the object. In particular, if the object starts at the position $x = 0$ then the position s at the time t is given by

$$s = ut + \frac{1}{2}at^2.$$

This is a quadratic equation linking position to time, with many major implications. For example, suppose that we know the braking force of a car; the formula allows us to calculate either how far we travel in a time t, or conversely, solving for t, how long it takes to travel a given distance. The notion of the *stopping distance* of a car travelling at a given velocity u results directly. Suppose that a car is travelling at such a speed, and you apply the brakes; how far will you travel before stopping? If a constant *deceleration* $-a$ is applied to slow a car down from speed u to zero, then solving for t and substituting the result in s gives the stopping distance $s = u^2/2a$.

This result predicts that doubling your speed quadruples your stopping distance. In this quadratic expression we see stark evidence as to why we should slow down in urban areas, as a small reduction in speed leads to a much larger reduction in stopping distance. Solving the quadratic equation correctly here could, quite literally, save your, or someone else's, life!

Source

An earlier version of this article appeared as '101 uses of a quadratic equation' in *Plus Magazine* (<http://plus.maths.org>) on 1 March 2004.

CHAPTER 10

Tony Hilton Royle Skyrme

ALAN CHAMPNEYS

The history of science is full of unsung heroes and forgotten figures. Sometimes, if the individual is lucky, due recognition comes in their own lifetime; sometimes it does not. For mathematicians it is especially unusual that one lives to see the true significance of one's work. A notable exception is the case of the recent Nobel Laureate Peter Higgs, Emeritus Professor of Theoretical Physics at the University of Edinburgh, whose name has recently been brought to the public attention. His theoretical calculations in the 1960s led to the prediction of a certain kind of subatomic particle, a boson, that bears his name. The probable discovery of the Higgs boson at CERN in 2012 has been hailed as the final piece of the jigsaw that confirms a remarkable mathematical theory, the so-called *standard model* of particle physics.

This confirmation of the standard model at the beginning of the 21st-century will doubtlessly contribute to immeasurable technological developments – in information technology, energy, and material science, for example – just like the first splitting of the atom at the beginning of the 20th century led eventually to many things we now take for granted. However, I can't help feeling that the media have allowed the Higgs boson to be enveloped in an unprecedented level of spin. Spin, not in the particle physics sense, but in an altogether less scientific, 21st century meaning of the word. As the TV character Jim Royle (no relation to the title subject's third name) of the BBC sitcom *The Royle Family* might say, 'god particle? my a**e!'

This article concerns another brilliant 20th-century British theoretical physicist, Tony Skyrme (Fig. 10.1). His claim to fame is the mathematical prediction of the *skyrmion*, a sort of distant cousin of the Higgs boson. Unlike Higgs, though, Skyrme died too early to achieve the recognition he deserved in his own lifetime.

Skyrme was a man I barely knew, yet I attended his funeral after he died suddenly in June 1987 following routine surgery, from what turned out to be an undiagnosed bleeding stomach ulcer. During my second year of undergraduate studies in mathematics at Birmingham University there were a number of post-exam lecture courses on optional topics of interest to current staff. A rather dishevelled figure that time seemed to have forgotten, Skyrme mumbled his way through a unique

Fig 10.1 Tony Skyrme (1922–1987).

series of fascinating lectures on the mathematics of symmetry, notes typed out in full on an old-fashioned typewriter. The lectures remained unfinished that year.

Tony Hilton Royle Skyrme was born in Lewisham in South London in 1922. After a scholarship to Eton, he took a first at Cambridge in mathematics, where he was president of the Archimedians, the world-famous mathematical society of undergraduate Cambridge mathematicians. In 1943 he completed Part III of the Mathematical Tripos, the unique extra year of mathematical studies offered by Cambridge University. He was immediately drafted into the war effort towards development of an atomic bomb. He worked under a former student of Werner Heisenberg, Rudolf, later Sir Rudolf, Peierls, a Jewish escapee from Nazi Germany. Their efforts were soon merged into the Manhattan Project in New York, before moving in 1944 to Los Alamos. Among other things, Skyrme worked on calculations using arrays of early IBM punch card business machines to predict the implosions needed to detonate a plutonium device.

Much of his Los Alamos work was classified; nevertheless, it formed the basis of his fellowship dissertation (essentially a PhD) at Cambridge in 1946. He took his fellowship to the University of Birmingham to work with Peierls, who was Professor of Mathematical Physics there. It was at Birmingham that Skyrme met his wife Dorothy, a lecturer in experimental nuclear physics in the University. Following temporary positions at MIT and Princeton, in 1950 they were appointed to research posts at the UK's Atomic Energy Research Establishment at Harwell.

Skyrme was by now a highly productive researcher and during the next ten years he produced a string of original papers on the fundamentals of subatomic physics. The contribution for which Skyrme is most famous is the introduction of the so-called Skyrme model, which described protons and neutrons in the context of recent theories of theoretical physics. His key insight was to pose a certain differential equation (one that essentially balances rates of change of physical quantities in space and time) that went beyond the existing linear theories. That is, Skyrme's model contained terms that went beyond strict proportionality.

The trouble was that, in the 1950s, there were no general methods for solving such equations. Solutions were expected to be mathematical descriptions of waves, metaphorically akin to the waves you can send down a rubber band held extended between your fingers. Through clever techniques, Skyrme was able to find a specific solution of a different type, corresponding, if you like, to a rubber band with a twist in it. Like the twist of a Möbius strip, such windings are stuck, so-called topological invariants, and cannot be removed simply by pushing or pulling. What Skyrme realised is that these kinds of mathematical objects are particle-like; the twist is localised in space and cannot be removed.

Starting in September 1958 Skyrme and his wife took a year's unpaid leave, during which they literally drove around the world. After a semester teaching at the University of Pennsylvania, they drove across America, taking in a nostalgic return visit to Los Alamos, before crossing the Pacific by boat to arrive in Sydney, Australia. Here, having purchased a Land Rover, and after spending a month lecturing at the Australian National University in Canberra, they drove through the heart of the red continent to Darwin in the north. Then they took a boat to Kuala Lumpur in Malaysia and literally fell in love with the place. Since they were both keen gardeners, they were taken by the lush tropical paradise. They then bought a Hillman and drove across Burma, India, Pakistan, and back to the UK via Iran, where Skyrme nearly died from salmonella.

On return to Harwell, Skyrme made plans to relocate to Kuala Lumpur, taking up a position in 1962 at the University of Malaya, where he became acting Head of the Department of Mathematics. He and Dorothy purchased a long-base Land Rover for the overland trip. As a practical man, he learnt how to strip it completely, and they carried with them a large stock of spare parts. The initial appointment was for three years, but it would seem all was not perfect in paradise and work became beset by administrative problems, not Skyrme's forte. So, when Peierls left for Cambridge in 1964, the position of Professor of Mathematical Physics at the University of Birmingham became vacant and Skyrme was the logical candidate. Skyrme remained at Birmingham until his death.

Skyrme was a shy man. He also believed that theoretical studies should be supplemented by physical work. In the 1950s he built his own TV and hi-fi. At Birmingham, he and Dorothy grew vegetables in their large garden, trying their best to be self-sufficient. Perhaps he feared the nuclear winter that might result from the atomic bomb that he had had a hand in developing. They had no children. Skyrme preferred to do his scientific work alone. He filled copious notebooks, but without the guiding influence of senior colleagues like Peierls or the stimulating atmosphere at Harwell, he saw no imperative to publish. He became somewhat reclusive, a rather forgotten figure, and would refuse invitations to speak at international conferences. Rob Curtis, my tutor at Birmingham, who would go on to become Head of Department, remembers Skyrme as a kind man who 'was always keen to teach as many courses as he was allowed, a refreshing contrast in these days when so many make strenuous efforts to have their teaching "bought out". In that respect, he was of the old school, genuinely "professing his discipline".'

It was not until the mantra of nonlinear science burst onto the scene in the early 1980s that Skyrme's work on nuclear physics began to get widely recognised. The name *skyrmion* for the localised states that Skyrme discovered was first coined at a small conference held in his honour in 1984 near Harwell. There is no formal record of that meeting. Three years later, though, Skyrme was persuaded to give a talk on 'The origins of skyrmions' at a celebratory meeting held for Peierls' 80th birthday. Tony Skyrme died two days before he was due to speak.

If Skyrme had lived, he would have seen a massive acceleration in interest in skyrmions. The skyrmion is a natural counterpart to the soliton, the kind of isolated wave that has been found to underlie many scientific phenomena from optical communications to tsunamis to structural buckling. A quick trawl through a popular scientific search engine reveals that up until 1987 there were about 100 scientific journal papers that mention skyrmions in the title or abstract. By the end of 2012, there were 1600 such papers, with 120 appearing in 2012 alone. Aside from their use in explaining the origins of certain subatomic particles, as an editorial in *Nature* in 2010 spelled out, skyrmion-like objects have been reported to be present in superconductors, thin magnetic films, and liquid crystals.

Tony Skyrme was never made a Fellow of the Royal Society, an accolade that most UK scientists of his standing should expect. Alfred Goldhaber quotes Morris Pryce, Professor of Physics at Oxford, that 'Tony was too clever for his own good'. Notwithstanding, in 1985 he was awarded the prestigious Hughes Medal by that same Society, one of the top international awards in the physical sciences and far rarer than being a mere Fellow. Dick Dalitz recalls from the award ceremony, 'As Tony came back to his seat, grasping his medal and with a grin all over his face, he muttered with justice, "They none of them could understand any of it at the time."' Interestingly, just four years previously the joint winner of the Hughes Medal was one Peter Higgs, for work leading to his prediction of the so-called god particle.

• •

FURTHER READING

[1] Gerald Brown (Ed.) (1994). *Selected papers with commentary of Tony Hilton Royle Skyrme*. World Scientific Series in 20th Century Physics, vol. 3. World Scientific.

[2] Dick Dalitz (1988). An outline of the life and work of Tony Hilton Royle Skyrme (1922–1987). *International Journal of Modern Physics A*, vol. 3, pp. 2719–2744.

[3] Alfred Goldhaber (1988). Obituary of T. H. R. Skyrme. *Nuclear Physics A*, vol. 487, pp. R1–R3.

CHAPTER 11

The mathematics of obesity

CARSON C. CHOW

From 1975 to 2005 the average weight of the adult American population increased by almost 10 kilograms*. The proportion that were obese increased by fifty per cent and this trend of increasing obesity could be found in most of the developed world. The obvious question is why. Why did the average weight remain relatively stable for most of the 20th century and then increase? There is no consensus about the cause. However, mathematics can help point to the answer.

To understand the obesity epidemic, we need to know what causes the human body to change weight. A direct approach is to perform a controlled experiment where people are randomly assigned to classes of diets and lifestyles and then monitored to see what causes weight gain. However, aside from the difficulties of tracking everything a person eats and does for several years if not decades, one must also ensure that their diets and lifestyles remain consistent throughout the trial. Such an experiment has never been fully implemented, and partial attempts to do so have provided inconclusive results. An alternative strategy is to use mathematics. A mathematical model of weight gain based on the underlying biology could answer the question directly.

There has been a long history of using mathematics in biology, with varying degrees of success. The main problem is that biology is exceedingly complex. The human body is an intricate collection of interacting organs, cells, and molecules, behaving in ways we do not fully understand. An attempt to create a mathematical model of a fully functioning human living in a realistic environment is well beyond our current capabilities. However, a well-worn strategy of applied mathematics is to attempt to isolate what is truly important for a phenomenon and create a simpler model. There is no guarantee that such an approach will work for every problem, but fortunately for weight gain we can rely on some physical principles to guide the construction of a mathematical model. The first thing to note is that we are effectively heat engines and are constrained by the laws of thermodynamics. The food we eat and the tissues in our bodies contain chemical energy that is stored in molecular bonds. The energy is released when fuel is burned or, more precisely, oxidised. In principle, you could heat your house by burning peanuts, potatoes, or, ghastly as it sounds, human flesh. We acquire energy in the form of food to maintain our bodily functions and to move about. According to the first law of thermodynamics, energy is conserved, so there are only three things that can happen to the energy in the food we eat – burn it, excrete it, or store it. If we can quantify how much energy we eat, burn, and excrete, we can compute how much energy we will store, and hence how much weight we will gain. Our goal is

---

*Strictly speaking, kilograms (or kilogrammes) measure mass rather than weight, but here common parlance is adopted.

thus to construct a mathematical model that balances the rate of change in body weight with the rate of food intake, the rate of energy excretion, and the rate of energy expenditure. Our bodies are generally quite efficient, so we do not excrete very much energy.

The problem is made slightly more complicated by the fact that not all of our tissues contain the same amount of energy. The primary energy-containing constituents of food and flesh are the macronutrients fat, protein, and carbohydrates. A gram of fat contains about 38 kJ (kilojoules) of energy, while a gram of protein or carbohydrates contains about 17 kJ. Usually, you eat food that consists of all three of these macronutrients. If you are in energy balance, you will burn or excrete everything you eat. However, if you are out of balance then you will either store the excess energy or make up for the deficit by losing some tissue by burning it. In the case of excess food intake, your body must then decide how to partition the extra energy into the various different types of tissue. While this process could have been very complex, nature was kind by providing a simple rule for partitioning energy; the ratio of the increase in fat tissue to all the non-fat or lean components of the body is proportional to the total amount of body fat. Thus in a cruel twist of fate, the more fat you have, the more fat you will gain if you overeat.

To complete the model, we need to know how much energy a person burns daily. This can be determined by measuring how much oxygen is consumed compared to how much carbon dioxide is expired. Experiments show that to a good approximation, the energy expenditure rate can be specified by a formula that is linear in the fat and lean components of the body. In other words, we can write the energy expenditure rate as $E = aF + bL + c$, where F and L are the masses of the fat and lean components, respectively, and the constants a, b, and c can be measured experimentally.

This ultimately leads to a mathematical model for the rates of change of the fat component F and the lean component L over time, written as dF/dt and dL/dt, respectively. In its simplest form the model obeys a system of two *coupled ordinary differential equations*

$$\rho_F \frac{dF}{dt} = (1 - p)(I - E),$$

$$\rho_L \frac{dL}{dt} = p(I - E),$$

where E is the energy expenditure rate, which is a linear function of F and L; ρ_L is the energy density of non-fat tissue; ρ_F is the energy density of fat; p is a function of F; and I is the intake rate of usable food, which is less than the total food intake rate owing to inefficiencies in the body's processing of material. Your body weight is given by the sum of F and L. All the parameters in the model can be specified by experiments, so this system of equations gives us the complete time-dependent history of the fat and lean components given the food intake rate. In other words, if I know how much you eat, I can tell you how much you will weigh. Changing the amount of physical activity you have will change the energy expenditure rate, and this too can be fully quantified.

It is possible to classify the types of dynamics that can arise in such systems of equations completely. Generally, such systems have *attractors*: as time evolves, values of F and L converge to a specific set of values that make up the attractor. Attractors might be *steady states*, in which F and L don't change over time, or periodic solutions with regular oscillations between a set of values, or more complex or even chaotic oscillations, though this latter possibility cannot occur in systems like ours with only two variables. In fact, it can be proved that the attractor here is a 1-dimensional curve of steady states. So there is not just one isolated steady state but a continuum of them, and solutions started from different initial conditions may tend to different steady states on the curve. This has the unsettling implication that two identical twins, eating the same diet and living the same lifestyle, could be in complete energy balance but have different amounts of fat and body weights depending on some quirk in their life histories.

The existence of this 1-dimensional attractor has another important consequence: it means that for most people, these equations can be approximated by something much simpler: a first-order linear differential equation of the form

$$\rho \frac{\mathrm{d}M}{\mathrm{d}t} = I - \epsilon (M - M_0),$$

where M is your current body weight, M_0 is some reference initial weight, ρ is an effective energy density of body tissue, and I is your total (effective) food intake rate.

Now, this simple equation tells us many useful things. For one thing, it can be exactly solved to give your weight for any diet. For an approximately constant daily food intake, the equation predicts that if you change your diet or lifestyle then your weight M will relax to a new steady state. The amount of time to relax, called the *time constant*, is given by the ratio of ρ to ϵ and for the average person is on the order of a year. Hence, if you change the daily food consumption rate from one fixed number to another, it will take about three years to get 95% of the way to your new steady state.

The second thing is that we can directly quantify how much your weight will change given how much you changed your food intake. The steady state is given by $M = M_0 + I/\epsilon$ (the value of M for which $\mathrm{d}M/\mathrm{d}t = 0$). For the average person, $\epsilon = 100$ kJ/kg. Thus for every 100 kilojoules change in your diet, your weight will change by a kilogram and it will take about three years to get there.

Finally, using this simple formula, we can show that an average increase of just 1000 kJ in daily intake for the American population between 1975 and 2005 explains the observed 10 kg increase in body weight. Agricultural data show that the amount of food available per person per day was approximately constant for most of the 20th century but then increased steadily from 1975 to 2005 by about 3000 kJ. Analysing the data with the mathematical model has demonstrated that the increase in food supply was sufficient to explain the obesity epidemic. In fact, the model predicted that if Americans had consumed all the extra available food, they would have been even heavier than they actually were. This implied further that the increase in obesity was accompanied by a concomitant increase in food waste, which was validated by direct data on food waste.

CHAPTER 12

It's a small world really

TONY CRILLY

I was sitting in an airport lounge waiting for a flight to Los Angeles when a fellow traveller struck up a conversation. He was from Iowa, which I barely knew, and I lived in London, but it was not long before we established common ground. It turned out that he had a friend whose cousin lived in New York, and this cousin was married to a person from London. It further emerged that the father of the Londoner lived next door to an uncle of mine and I had met him over a cup of coffee only a couple of months earlier. As the flight was called and we made our way to the departure gate there was only one possible farewell: 'small world'.

The connections between ourselves and people met at random seem nothing short of miraculous. But I wondered if they really were so surprising.

We can well imagine the population of the world with lines of connection attached between pairs of individuals making up one gigantic network, rather like a map of the globe showing airline routes. In my airport conversation the connections were of various types: friendship, blood relationships, and living in the same location.

We can also imagine other 'worlds'. At one extreme we can think of a population where everybody knows everybody else, in which case the world would be a morass of connecting lines going every which way. In this world two random people would be directly connected with each other. At the other extreme is the world where no two people are connected, and in this world there would be a total absence of connections. In the real world neither of these extremes exist: some people are connected with many people, some with a few, and others connected with none. Two people may be connected with an intermediary, like A is connected to B and B is connected to C so that A is connected to C through the intermediary B. It is also possible for there to be several intermediaries.

A real-life situation for alternative worlds would be the members of Facebook, in which each individual has a set of friends. Some people have hundreds of friends, in which case they are called *hubs*, but others restrict their friends to their immediate family and are comparatively isolated. I wondered about the New York cousin of my Iowa acquaintance. Was she one of these hubs in the chain which linked us, the chain *him–friend–cousin–husband–father–uncle–me*? This was a sixfold separation, six being the number of dashes in the chain.

One question kept reoccurring. Is the existence of such connecting chains so surprising? This is where mathematics steps in and helps to explain the phenomenon of such a close connection between random individuals.

As with most things mathematical the story goes back to concepts learned centuries ago, this time to the idea of logarithms. This theory was developed in the early 1600s by such people as John Napier and Henry Briggs but for completely different purposes. In those days the multiplication of large numbers presented a practical problem, for such calculations could be both laborious and time-consuming.

So the brilliant idea of our forebears was to change the numbers into logarithms using a mathematical table, then simply add the logarithms, and reconvert the result backwards to get the answer to the multiplication sum. At a stroke, multiplication was replaced by addition, a much simpler operation. The word 'logarithm' is derived from the Greek and is based on ratio and number, but we don't have to know that to make sense of logarithms.

Logarithms are easiest to compute for powers of 10, and we naturally focus on 10 because it is the basis for our number system. The common logarithm of 100 is 2 because $100 = 10 \times 10 = 10^2$; the common logarithm of 1000 is 3 because $1000 = 10 \times 10 \times 10 = 10^3$. We can shortcut this by noting that the logarithm of these numbers is the number of zeros and we write $\log(100) = 2$ and $\log(1000) = 3$, and similarly $\log 10 = 1$. If we wanted to know the logarithm of 536 (say) we would have to find this out from logarithm tables, though as 536 is between 100 and 1000 we already know its logarithm would be between 2 and 3. In fact, $\log(536) = 2.7482$.

This theory of logarithms is just what we need to understand the phenomenon of chains connecting individuals. The main mathematical result was proved by American mathematicians Duncan Watts and Steven Strogatz in a paper published in the journal *Nature* in 1998, a paper which was one of the most cited of any dealing with scientific topics. They proved that if a population is of size P, where each person has K direct but randomly distributed acquaintances, then the average number of degrees of separation is the logarithm of the population divided by the logarithm of the number of connections; in symbols,

$$\frac{\log(P)}{\log(K)}.$$

To see how this works we might consider a small town like Bideford in the United Kingdom, with a current population of about 15,000 people. So in this case $P = 15,000$, and we will assume each person has (say) about $K = 20$ acquaintances. From the mathematical result we calculate

$$\frac{\log(15000)}{\log(20)}$$

and the calculator gives this is as

$$\frac{4.1761}{1.3010},$$

which is 3.2099. This means that between any two people in Bideford we would expect there to be three degrees of separation. That is, if strangers A and B are sitting in a coffee shop in Bideford and strike up a conversation, we would expect them to discover they have two intermediate acquaintances C, D so as to form a chain $A–C–D–B$. There are three degrees of separation between A and B.

If we focus on the population of the Earth, currently 7 billion (and then discounting 20% as an estimate of the number with zero connections), this gives us a working number of $P = 5,600,000,000$ people and, with on average $K = 40$ acquaintances (say), the average number of degrees of separation can be calculated according to the paper by Watts and Strogatz as

$$\frac{\log(5,600,000,000)}{\log(40)}.$$

From tables this is

$$\frac{9.7482}{1.6021} = 6.0846,$$

which means there are on average 6 degrees of separation between two individuals.

This number depends on the number of direct connections we assume. But if we repeat the calculation for a number of direct connections K in the range between 25 and 90, the average number of degrees of separation only varies between 7 and 5.

In fact, Watts and Strogatz showed that if some of the people had long-range connections then the number of degrees of separation could be smaller, often much smaller. For example, if a person in Bideford knows someone in New York, then a friend of that person would also have a connection to people in New York. Watts and Strogatz called networks with mostly short-range connections, but with the occasional long-range connection, *small world networks* (see Fig. 12.1). These networks are a good representation of the way that surprising connections are made in real life.

This theory of separation is the basis of the game called Six Degrees of Kevin Bacon (Fig. 12.2). Kevin Bacon is a Hollywood actor who has appeared in many films. Another actor is considered to be directly linked to him if they appeared in a film together. But then another actor D may appear in a film with C who has appeared in a film with Kevin Bacon. In this case D is connected to Bacon by the chain $D–C–B$, in which case D is said to have Bacon index two, corresponding to the two degrees of separation. Hollywood aside, mathematicians take pride in an *Erdős index*, which is constructed similarly. What is an Erdős index? A person has an Erdős index of two if that person has written a mathematical paper with someone who has written a paper with Paul Erdős. This prolific researcher toured the world knocking on the doors of mathematicians, staying over in their houses, writing a paper with each of them, and moving on. He cared little for material possessions and gave most of his money away, but he had an extraordinary passion for mathematics.

So we find coincidences are not so coincidental as we might at first think. The newsworthy *six degrees* is actually a property of large networks of random connections, and we notice that statements are couched in terms of the average. It might be the case that two individuals, plucked at random, have very high degrees of separation. It's just that the theory tells us an average number of degrees and this is six.

The mathematics of connections is usually regarded as part of *graph theory*. The work on the 'six degrees of separation' is a subsection of this theory and is allied to *random graph theory*. Graph theory itself can be traced back to the 18th century and the work of the most prolific mathematician of all time, Leonhard Euler, while random graph theory is a more recent study. It is an energetically researched subject today which has gained importance due to its applicability and usefulness in computer science. It also says something useful about human connections. Next time you are in an airport lounge or sitting in a coffee shop, you may look around and muse on the mathematical theory making ours a really small world.

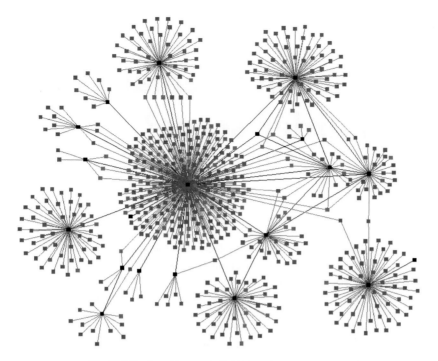

Fig 12.1 Small world network. © AJ Cann, Creative Commons.

Fig 12.2 A small subset of the 'six degrees of Kevin Bacon'. © Philipp Lenssen, Creative Commons.

CHAPTER 13

How does mathematics help at a murder scene?

GRAHAM DIVALL

Blood is shed during many crimes of violence, and the examination and characterisation of bloodstains is an important part of a forensic scientist's work. Before using powerful identification techniques, such as DNA profiling, a great deal of useful information can be obtained from the nature, shape, and distribution of bloodstains at the scene of a crime.

Imagine the scene of a violent murder where the victim has been bludgeoned to death. The scene will contain many bloodstains of different shapes and sizes. Some will be smears, others will be spots, and many will be in the form of splashes; see Fig. 13.1 for a representation. Each of these stains contains a hidden message that can tell how the bloodstain was made. Mathematics can be used to read these messages and contribute to reconstructing the crime scene.

Blood can drip passively from a bleeding wound or from a blood-soaked weapon. Under the force of gravity the drops fall vertically and when each drop strikes a horizontal surface it forms into a circular stain. The diameter of the stain and the appearance of its edge depend on the volume of the drop, the distance it has fallen, and the physical characteristics of the surface.

In contrast, many drops of blood are projected from their source or origin. This can occur, for example, when a bleeding injury is repeatedly struck with a fist or a hammer, or by blood being flicked off a weapon when it is swung through the air and changes direction. As shown in Fig. 13.1, these blood drops tend to hit a horizontal surface such as a floor at an acute angle θ and the stain formed is in the shape of an ellipse. The ellipse is characterised by the length of its major and

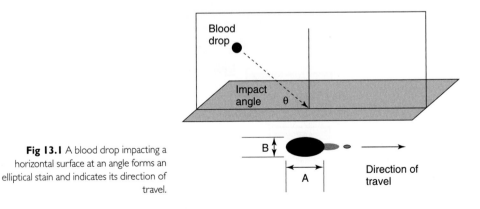

Fig 13.1 A blood drop impacting a horizontal surface at an angle forms an elliptical stain and indicates its direction of travel.

minor axes, A and B. As the impact angle θ decreases, A gets longer and B gets shorter. Visually, the bloodstain becomes a longer and thinner ellipse. Of great use in crime scene reconstruction is the relationship

$$\sin\theta = B/A.$$

This means that if we can measure A and B for a given stain, then we can calculate the angle θ at which the drop of blood hit the surface.

These blood splashes also contain another important piece of information, as shown in Fig. 13.1. Although most of the stain is in the shape of an ellipse, it tends to be elongated on one side of the ellipse's major axis and often has an additional small stain, known as a cast-off or satellite stain, on the same side. These features are important in crime scene reconstruction because the elongation and cast-off stain both point in the direction the drop of blood was travelling when it hit the surface.

Armed with information about impact angles and directionality, we can now determine a region in space from which a group of blood splashes could have originated. There are several methods for doing this, all of which use basic trigonometry. Figure 13.2 illustrates the principles using three stains.

First, for each stain in the group we draw backwards its line of directionality. If the stains have a common origin, then these lines will converge to a common point P. For each stain, the distance d from the stain to the common point is measured. Then, we assume the flight path or trajectory for each stain is a straight line and calculate the vertical distance h above the common point using $h = d\tan\theta$. This locates the origin of the stain in 3-dimensional space, and for all the stains these collectively form a region of common origin. This is not a point source, for several reasons. First there are errors associated with the measurement of the stains. Also, in reality, such blood splashes arise, say, from an injured head, which does not correspond to a point source.

Other methods for locating the region of common origin involve simple geometric construction. An old method required a length of string to be located on each stain, and this was then stretched in a straight line at the impact angle θ so that the line and the line of directionality were

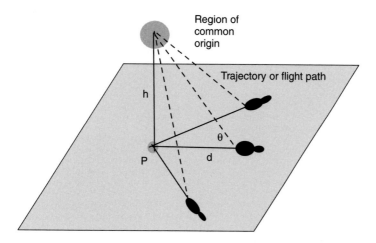

Fig 13.2 Locating the origin of a group of blood splashes.

in the same vertical plane. The strings from a group of stains would converge to the region of common origin. This cumbersome method has now been largely replaced by modern computer graphics software but the procedure is still known as 'stringing'. The same principles can be used to analyse blood splashes that are located on the ceiling of a room and on vertical surfaces such as walls and sides of furniture.

The procedures described help locate the origin of groups of blood splashes at the scene of a crime. They provide useful information such as the sites in a room where a victim was attacked and the number of blows a victim received.

We have seen that these methods model the formation of blood splashes at a crime scene and, like a lot of mathematical models, it is necessary to make certain simplifying assumptions. For example, we ignore the effect of air resistance on the flight of a blood drop. More importantly, the model assumes that the trajectory or flight path of a blood drop is a straight line. For the majority of drops, this is not the case. Once a drop of blood is projected from its source it is subject only to the force of gravity, and the drop follows a parabolic path until it hits a surface. This means that the model predicts a region of common origin that is higher than the actual origin. The problem is that we cannot predict the motion of blood drops as projectiles because we do not know important features such as launch velocity, impact velocity, and time of flight. All we have is the final bloodstain. But all is not lost. Scientists are now examining features of the final stain such as its size and edge characteristics in much greater detail and attempting to derive equations for the blood drop's impact velocity. It will be a few years yet, however, before these approaches can be used in real crime scene reconstruction.

. . . .

FURTHER READING

[1] Stuart James, Paul Kish, and Paulette Sutton (2005). *Principles of bloodstain pattern analysis*. CRC Press.

Source
Selected as an outstanding entry for the competition run by *Plus Magazine* (<http://plus.maths.org>).

CHAPTER 14

Mathematics: The language of the universe

MARCUS DU SAUTOY

Mathematics can often appear arcane, esoteric, unworldly, irrelevant. Its blue-sky status could easily make it a target for the harsh cuts that governments are contemplating to science budgets around the world. But before the axe falls it is worth remembering what the great 17th-century scientist Galileo Galilei once declared: 'The universe cannot be read until we have learnt the language and become familiar with the characters in which it is written. It is written in mathematical language, and the letters are triangles, circles and other geometrical figures, without which means it is humanly impossible to comprehend a single word.'

The scientists at CERN will certainly agree with Galileo. The reason they are able to make predictions about the particles they are expecting to see inside the Large Hadron Collider is entirely down to mathematics. Rather than triangles and circles it's strange symmetrical objects in multidimensional space, shapes that only exist in the mathematician's mind's eye, that are helping us to navigate the strange menagerie of particles that we see in these high-energy collisions. Biochemists trying to understand the 3-dimensional shapes of protein strings will also sympathise with Galileo's sentiments. Each protein consists of a string of chemicals which are like letters in a book. But to read this language and predict how the 1-dimensional string will fold up in 3-dimensional space, you need a mathematical dictionary. Studying protein folding is key to understanding several neurodegenerative diseases which are a result of the misfolding of certain protein strings.

Modern astronomers use mathematics to penetrate the night sky just as Galileo used the telescope to see things that were beyond the reach of the human eye. Instead of drawing the shapes of lions and bears through the chaotic smattering of stars, mathematics turns these lights into numbers that can then be read to find out where our universe came from and, ultimately, to predict what will be its future.

Although mathematics has offered science an extraordinary beacon with which to navigate the universe, it is striking how much of the subject still remains a mystery. How many mathematical stories are still without endings or read like texts that have yet to be deciphered?

Take the atoms of mathematics, the primes. The indivisible numbers like 7 and 17 are the building blocks of all numbers because every number is built by multiplying these primes together. They are like the hydrogen and oxygen of the world of mathematics. Mendeleev used the mathematical patterns he'd discovered in the chemical elements to create the periodic table, the most fundamental tool in chemistry. So powerful were these patterns that they helped chemists to predict new elements that were missing from the table. But mathematicians are still to have their Mendeleev moment. The pattern behind the primes which might help us to predict their location has yet to be uncovered. Reading through a list of primes is like staring at hieroglyphs. We have made progress and have unearthed something resembling the Rosetta stone of the primes. But decoding the stone still eludes us.

Mathematicians have been wrestling with the mystery of the primes for 2000 years. But some of the biggest problems of maths are far more recent. There was a flurry of excitement recently that the *P versus NP* problem had been cracked. First posed in the 1970s, this is a problem about complexity. There are many ways to state the problem, but the classic formulation is the *travelling salesman problem*.

An example is the following challenge: you are a salesman and need to visit 11 clients, each located in a different town. The towns are connected by roads, as shown in the map in Fig. 14.1, but you only have enough fuel for a journey of 238 miles.

The distance between towns is given by the number on the road joining them. Can you find a journey that lets you visit all 11 clients and then return home without running out of fuel? (The solution is at the end of the article.) The big mathematical question is whether there is a general algorithm or computer program that will produce the shortest path for any map you feed into the program that would be significantly quicker than getting the computer to carry out an exhaustive search (another version of this puzzle has the more restrictive stipulation that each road must be used once and only once, see Chapter 20). The number of possible journeys grows exponentially as you increase the number of cities, so an exhaustive search soon becomes practically impossible. The general feeling among mathematicians is that problems of this sort have an inbuilt complexity, which means that there won't be any clever way to find the solution. But proving that something doesn't exist is always a tough task. The recent excitement that this problem had been cracked has since evaporated and the problem still remains one of the toughest on the mathematical books.

But the recent solution of one of the most challenging problems on the mathematical books, the Poincaré conjecture, gives us hope that even the most elusive problems can be conquered.

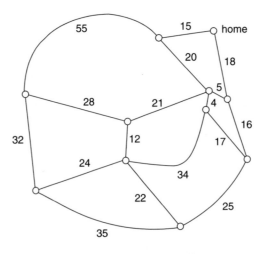

Fig 14.1 Travelling salesman problem.

The Poincaré conjecture is a fundamental problem about the nature of shape. It challenges mathematicians to list the possible shapes into which 3-dimensional space can be wrapped up. In 2003 the Russian mathematician Grigori Perelman succeeded in producing a periodic table of shapes from which all other such shapes can be built.

These fundamental mathematical questions are not just esoteric puzzles of interest solely to mathematicians. Given that we live in a world with three spatial dimensions, the Poincaré conjecture tells us ultimately what shape our Universe could be. Many questions of biology and chemistry can be reduced to versions of the travelling salesman problem where the challenge is to find the most efficient solution among a whole host of possibilities. A resolution of the P versus NP problem could therefore have significant real-world repercussions.

Modern Internet codes rely on properties of prime numbers. So any announcement of a breakthrough on the primes is likely to excite the interest not just of pure mathematicians but also of e-businesses and national security agencies. In a time when blue-skies research without obvious commercial benefits could be under threat from huge sweeping cuts it's worth remembering how Galileo concluded his statement about the language of mathematics: without it we will all be wandering around lost in a dark labyrinth.

Solution to travelling salesman problem:

$$15 + 55 + 28 + 12 + 24 + 35 + 25 + 17 + 4 + 5 + 18 = 238.$$

Source
This article first appeared in *The New Statesman* on 14 October 2010.

CHAPTER 15

The troublesome geometry of CAT scanning

RICHARD ELWES

In 1895 the scientist Wilhelm Röntgen became an unlikely celebrity when he took some remarkable photographs of his wife's hand. To widespread astonishment, these pictures revealed the very bones of her living fingers. Her wedding ring stood out too but, seemingly magically, her skin and flesh were rendered invisible. Röntgen's *X-rays* caused a media storm and went on to revolutionise physicists' understanding of electromagnetism. The consequences for medical science were no less profound and can be seen in hospitals around the world, with hundreds of millions of X-ray photographs now taken every single year, improving diagnoses and saving countless lives. But today's medics are not always content with flat 2-dimensional pictures of the kind that Röntgen first produced. For a simple broken arm, a conventional X-ray photograph may be adequate. But to understand the exact size and location of a cancerous tumour demands something more: a 3-dimensional map of the living body, inside and out. In the 1960s and 1970s the technology to do this arrived in the shape of *computed axial tomography* (CAT) machines (tomography being the science of scanning).

The road from Röntgen's plain X-ray photographs to today's CAT scanners involved several advances in physics and engineering. But it also required the solution to one particularly thorny geometrical problem, namely, how can we hope to reconstruct the patient's 3-dimensional structure from a collection of 2-dimensional images? As it happens, this problem was studied and solved by the Austrian mathematician Johann Radon in 1917, though at the time he had little notion of the deep practical implications of his work.

X-ray photographs

Back in the 19th century the newspapers hailed the discovery of X-rays with breathless excitement. Nowadays we know that they are not magic but a type of radiation, like visible light but at a shorter wavelength. In fact, medical X-rays clock in at a wavelength of around 0.01 nanometres, which is forty thousand times shorter than ordinary red light. It is this difference that gives X-rays their apparent superpowers: while visible light cannot permeate most solid objects (with a few exceptions such as glass), X-rays generally can.

However, if X-rays simply travelled straight through all matter unimpeded they would be of little medical use. It is critical that some proportion of the radiation is absorbed or scattered along the way. What is more, certain types of materials (such as bone) absorb more radiation than others (such as muscle and fat). It is this difference that produces an informative image on the photographic paper beneath, which becomes darker the higher the energy of the ray striking it.

Useful as they are, X-ray photographs of an object do not tell the whole story. To see why, imagine being presented with an X-ray image of some unknown object and being challenged to deduce its 3-dimensional structure. You might reason that if the image is bright white at some particular point, then most of the corresponding rays' energy must have been absorbed or scattered en route. The trouble is that the same result might be produced by a thin sheet of metal or a much thicker layer of something less absorbent, such as wood. There is no way, from a single picture, to tell the difference. The usefulness of traditional X-ray photographs in a medical setting relies on staff who already know the approximate 3-dimensional locations of the various bones and organs being imaged.

CAT scanners and cross-sections

CAT scanners can do better. The principle by which they work is essentially the same as the traditional X-ray photograph. The difference is that instead of a single image, the patient's body is photographed from multiple angles simultaneously (or nearly so). These various images are then assembled to give a 3-dimensional map of their body ... which is where some serious mathematics enters the frame.

What we are hoping for geometrically is a mathematical rule or *function* (call it F) that takes as input the (x, y, z)-coordinates of a point (call it p) in 3-dimensional space. The function should return as its output, $F(p)$, the level of absorption of the material at the point p. If p is a point within a muscle, then $F(p)$ will be a small number since X-rays can easily penetrate muscle. If p is somewhere within a bone then $F(p)$ will be much higher, and if it is within an artificial metal joint then $F(p)$ will be very high indeed. So knowing the value of F at every point p would tell us exactly what we want; it would provide a complete 3-dimensional map of the patient.

The first observation about the problem of finding such a 3-dimensional function F is that it really reduces to a series of 2-dimensional problems. In a modern CAT scanner, X-rays are fired in parallel from a series of emitters on one side of the patient's body towards detectors on the other. The arrays of emitters and receivers then each rotate 180° around the patient but, critically, each detector remains aligned with the same emitter, with all its rays passing through the same 2-dimensional cross-section of the patient's body. (The word 'axial' is where the 'A' in the definition of CAT comes from.) If the geometry of each cross-section can be understood thoroughly, it is then simply a matter of assembling them in the right order to obtain the required 3-dimensional map. So, the geometrical nub of the problem is to reconstruct a single 2-dimensional cross-section of the patient's body from the X-ray data.

The Radon transform

What we are hoping for is a map giving the X-ray absorption level at every point in a 2-dimensional object. This is the goal of the scanning process. The trouble is that this is not the

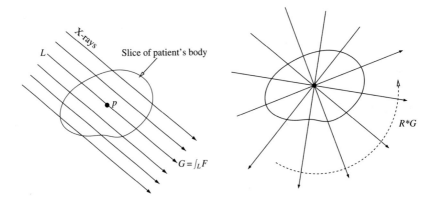

Fig 15.1 (Left) The definition of the Radon transform $G = RF$ and (right) its dual transform R^*G.

data that the scanner provides. At the end of an individual ray (call it L), the scanner will determine the *total* absorption of all the material through which that beam has passed. This amounts to a sort of sum of the absorption at each point along the straight line L. Mathematicians commonly denote this sort of continuous summation using the integral symbol, which looks a bit like an elongated capital S (for 'sum'). So we might might write this total as $\int_L F$.

So the scanner produces a new function related to but different from F. It is known as the *Radon transform* of F (Fig. 15.1). We'll denote it by RF (R for Radon). This function represents an input–output rule of a different kind: while F's inputs are the coordinates of a point p in space, with outputs of the absorption at this location, an input for RF is a line L, and its output is the total absorption along L, which is to say $\int_L F$.

Now, if we knew F for every point p it would be a simple matter to calculate RF for any line should we wish to know it. But the problem we are faced with is to go the other way: what the scanner tells us is the value of RF, for different lines L, while we want to know F for all points p. Mathematically then, the central question is whether there is some way to work backwards from the data we are given, to find an explicit description of what we really want, namely of F itself. Luckily, the answer, Johann Radon was able to show, is yes.

Radon inversion

Suppose that we scan a patient and again focus on a single cross-section of the scan. In essence, the resulting data amount to a function, G, that takes a line L as an input and outputs the total absorption along L. The mathematical challenge is to find the correct F such that $G = RF$. This problem is known as *Radon inversion* and is the central mathematical conundrum in tomography.

To solve this central problem, a good first step is to calculate the so-called *dual transform* of G. This time, instead of totalling along a particular line, we fix a point in space (say p), and average across all the lines that pass through p. We might write this, rather informally, as $\int_p G$. This describes a new function, which we might call R^*G, related to G, but which takes a point p as input and outputs the average absorption of all the lines that pass through p.

The number produced by R^*G is not yet the one we need. Unfortunately its value at p is not the absorption at that point but the average absorption of all the lines that pass through p. What we actually want is a kind of weighted version of this average so that points nearer to p contribute more than those further away. So one final mathematical step is needed.

Radon's original work on his eponymous inversion problem was a milestone in the subject of integral geometry. Since then, several other approaches to the problem have been developed. Most of them invoke the same idea in order to finish the computation and reduce F from R^*G. They use yet another transform, a superstar of the mathematical world, the *Fourier transform*, whose origins lie in the seemingly unrelated theory of waves.

Fourier transforms

The simplest, smoothest, and most symmetrical type of wave is the sine wave, which undulates with perfect left–right and up–down symmetry. But, when viewed on an oscilloscope, most physical waves, such as a human heartbeat or the sounds produced by musical instruments, have far more complex appearances. Yet in the early 19th century Joseph Fourier discovered the fundamental fact that every such waveform can be built from a suitable combination of sine waves. All it needs is for the sine wave at each frequency to be adjusted to the correct output level, or volume setting, if you like. The rule that assigns the appropriate volume level to each frequency is known as the Fourier transform of the original, more complicated waveform.

The Fourier transform is of incalculable value in the modern world, for example in technology such as sound-sampling, where physical waves are converted into digital signals. Indeed, inside every mobile communication device, there will be a computer chip carrying out Fourier transforms to recover signals that are embedded inside transmitted electromagnetic waves.

It does not seem especially promising to imagine the tissue of the patient being composed of waves, so one might wonder how the Fourier transform comes into play in CAT scanning. First, the Fourier transform has the advantage of being much easier to invert than the Radon transform. Indeed, the processes of calculating the Fourier transform and finding its inverse are almost identical. This elegant symmetry is the source of the Fourier transform's value. The second answer is that the two transforms, Fourier and Radon, interact in a particularly satisfactory way; after a small extra mathematical tweak, it turns out that the Fourier transform provides exactly the step needed to complete the calculation and reconstruct F from R^*G.

Although the procedure of computing R^*G and then taking a Fourier transform is somewhat complex, Allan McCleod Cormak established in 1963 that it can be carried out systematically and algorithmically. Nevertheless, after the first medical CAT scan in 1971, the computation to reconstruct the image lasted 2.5 hours. But with the power of modern computing and algorithms that have improved over several generations, the calculations can now be completed almost instantly. Indeed, this is precisely what happens in hospitals around the world, every time a patient slides into a medical CAT scanner.

• •

FURTHER READING

[1] Richard Elwes (2013). *Chaotic fishponds and mirror universes*. Quercus.
[2] Chris Budd and Cathryn Mitchell (2008). Saving lives: The mathematics of tomography. *Plus Magazine*, <http://plus.maths.org/content/saving-lives-mathematics-tomography>.

CHAPTER 16

The mathematics of sports gambling

ALISTAIR FITT

L et's get one thing straight first – this article probably won't tell you what you want to know. You won't learn how to get rich quick, you won't hear about the huge fortune that I have accumulated from gambling (there is none), and you won't hear about my plans to give up my day job and become a professional gambler (though I don't rule this one out completely). You *will* hear about what the mathematics of gambling involves, how you might use your mathematical knowledge to gain an edge, and the kind of issues that all of us who are currently interested in 'intelligent' gambling are currently thinking about.

You might of course also be tempted to try out some of the strategies that are hinted at in this brief article, so I make no apologies for starting with my normal warning:

The advice given in this article is the author's personal view and may be utterly misguided, so if you stake even one solitary penny based on what you read here, then the responsibility is entirely yours and the author (much less the IMA) will not be held responsible for your losses.

Punters, predictors, and arbers

Who's gambling today? Well, largely, there are punters, and there are mathematical gamblers. No further comment will be passed on punters, who bet largely for pleasure, lose small amounts most of the time, and often have selective memories. (We don't disapprove of punting – it can be innocent fun and it can also be unhealthily addictive. It's simply not of interest here.) Mathematical gamblers (like us) bet for gain, and are uninterested in the underlying event. We bet when the circumstances are right to make money, not when we fancy a flutter.

Mathematical gamblers essentially divide into two classes: predictors and arbitragers (*arbers*).

Predictors

The one constant star of predictors is that they don't back the selection (for convenience let's call it a horse – it could be anything) that they think will win. Instead, predictors back the horse that is advantageously incorrectly priced. Much of the time this will lead to a loss, but if the strategy is faithfully followed and only *expected win* bets (see below) are ever made, an overall profit is assured. Operating in this fashion normally requires a great deal of data, modelling, and patience. It may also require a large, constantly changing support team to actually place the bets. A very few people make big profits this way. (See the further reading suggestion below – you can decide how much of it to believe. You might also find it interesting to Google the name 'Bill Benter'.)

Arbers

In contrast to predictors, arbers are completely uninterested in the race result and never seek to identify the winner. Instead, they (let me be upfront – we) seek circumstances (see below for examples) where the odds have a particular structure that allows a guaranteed win whatever the result of the race. In short, we only ever bet if we know that we are going to win. It turns out that some arbitrages are very simple but essentially useless, while the useful ones are often extremely complicated – that's where the mathematics comes in. Note: at this stage everybody always asks, 'if arbitrage is possible, why aren't you a millionaire?' Some of the answers to this very reasonable question will be addressed below.

Simple betting theory and *expected win* bets

To understand what's going on, a few definitions are unavoidable. First, we use so-called *decimal odds* throughout. When we 'back a horse at 4.5' (say) we mean that we stake £1 at decimal odds of 4.5. If the horse wins, our profit is £3.5 (£4.5 less the stake); if it loses, our profit is –£1 (a loss). Suppose we denote the probability of an event X by $\text{Prob}(X)$. Then an *expected win* bet (the vital raw material of all predictors) is simply one where the expected winnings are positive and the quantity

$$(\text{Prob}(win) \times (\text{winning profit}) - \text{Prob}(lose) \times \text{stake})$$

exceeds zero.

How could an *arb* work? Suppose in a two-horse race the two horses have odds W_1 and W_2. Then it's not hard to see that if

$$R = \frac{1}{W_1} + \frac{1}{W_2} < 1$$

then betting respective amounts $\lambda_1 = P/(W_1(1 - R))$ and $\lambda_2 = P/(W_2(1 - R))$ on the horses yields a guaranteed profit of P whatever the result of the race (if horse 1 wins then we win $\lambda_1(W_1 - 1)$ and lose λ_2: a similar sum also gives a profit P if horse 2 wins). Analogous results and formulae may easily be derived for an N-horse race with little effort. In reality, this never of course happens. The quantity R (the 'roundness of the book') invariably exceeds 1, meaning that an arbitrage of this sort is not possible. Nevertheless, as we shall see below, this does not mean that arbs are impossible.

Arbing and the exchanges

Who do we bet with? Not bookmakers. Most of us who are interested in betting have had on-line accounts with High Street bookmakers. None of these have lasted very long, though, for the bookmakers' rules are simple: if they detect you doing 'smart' things, they will either close your account or gag you by only allowing very small bets. This was a huge problem for all serious sports gamblers until 2000, when the first betting exchange launched. Betting exchanges such as Betdaq and Betfair are essentially 'eBay for bets'. No bookmakers are involved, the great betting public bet with each other, and a computer instantly matches the bets (very much like the London Stock Exchange).

Betting exchanges make their profit by taking a few per cent of each winning bet (though they generously allow you to lose commission-free). The other great innovation introduced by the exchanges was the ability to lay selections ('bet on a horse NOT to win') as well as back them. A lay bet works in the opposite way to a back bet. For example, if a horse's decimal lay odds are 9.4 then a £2 lay bet wins a profit of £2 if the horse fails to win, but loses £16.80 (= £2 × (9.4 − 1)) if the miserable nag should happen to triumph.

The simplest arb

The simplest arb that I know occurs maybe a few times a day. Bookmakers allow 'each-way' bets on most races, where equal stakes are backed for a horse to win and to place. The place yields a fixed fraction of the odds (for example, in a six-horse race 1/4 of the odds is paid for the first two places). By solving a simple linear system of 3 × 3 equations where the three unknowns are the each-way bet, the win lay, and the place lay, one can determine circumstances where backing a horse each way with a bookmaker and laying both the place and the win on an exchange (at the correct stakes) will yield a guaranteed profit. For example, on 8 March 2013 in the six-horse 19.10 at Wolverhampton, Assizes could be backed each way with a bookmaker at 5.00, and the win and place laid on an exchange at 5.08 and 1.75, respectively (the details are left as an exercise). Unfortunately, as explained above, bookmakers will soon spot what you are doing and eject you.

The arb challenge

Since the simplest arb is essentially useless, how can we proceed? The answer is that other arbs are available that use only exchange markets. Your imagination is the only limit to designing arbs for various sporting bets that are traded on the popular betting exchanges. Tennis (which typically attracts very large, varied, and liquid markets) is a particularly rich source of possibilities, but it's not hard to construct other arbs made up of combinations of bets. The challenge is to find arbs that have high utility (i.e. the profit ratio is high, so one is not limited by the amounts available to back or lay) and occur often enough to be useful.

Battle of the bots

You have probably gathered by now that arbing normally involves placing a large number of bets simultaneously with carefully controlled stakes, at an instant when the price is exactly right. Inevitably, this cannot be done manually, so we use robots ('bots') to place bets. Most exchanges encourage bot betting by making bot-writing software available. A typical 'arb attack' might consist of the following steps:

- 'scrape' a betting exchange site (using, for example, a PYTHON/BeautifulSoup script) to identify circumstances where an arb is available;
- while the scrape is on, calculate the bets required according to formulae that we have developed and coded;
- use bots to place the required bets on various exchange markets.

What do we conclude? Simply that it is easily possible (using some fairly basic mathematics) to make a risk-free profit from a betting exchange. The main challenges include (i) one's ability to automate the processes involved, (ii) finding arbs that occur sufficiently often and have sufficient utility, and (iii) finding schemes where the profits available are not too limited. Most likely you will not become rich, but you can certainly have a lot of fun and make some spare cash into the bargain. Space constraints have prevented me from discussing the more complicated mathematical aspects of arbitrages and expected win bets, but some of the mathematical challenges are undoubtedly interesting and could lead to worthwhile publications (providing of course that one remembers the golden rule of betting research, namely to only publish good ideas that don't actually work).

. . . .

FURTHER READING

[1] Patrick Veitch (2009). *Enemy number one: The secrets of the UK's most feared professional punter.* Racing Post.

PYTHAGORAS'S THEOREM: a^2

Hommage à Queneau

Bare bones

The square on the hypotenuse equals the sum of the squares on the other two sides.

Chinese whispers

The square hippopotamus has a tum in which his other tooth hides.

Modern art

The Triangle is one of the most influential examples of early squarist art. Created by the immortal artist Pythagaro, its stark simplicity has much to commend it to more modern eyes. Its three sides signify the fusion of Art, Beauty, and Creation. The squares on each side expand these three to combine them with the opposing elements of air, earth, water, and fire to create a hypotenic mixture of angle and area. Each square is linked to us as humans by being filled with many small squares each with an image of a friend of the artist. By an extraordinary piece of artistic genius, the number of friends in the Art and Beauty squares exactly matches those in the square of Creation. Such depth of meaning shows us an immortal truth – priceless!

Banach–Tarski

The square on the hypotenuse can be cut up and reassembled to create two squares of any required area provided non-measurable partitioning is allowed.

Personals

Fun loving Greek philosopher with good sense of geometry seeks meaningful relationship and would consider being in a triple. Nothing irrational please. Apply: Box 345.

Management consultant

Given the right core competencies we must triangulate our key deliverables by close of play. The message going forward is to start at the basics and avoid the linear relationships between any two points in our supply chain. We must keep our eye on the prize, go both wide and deep, and focus on ballpark estimates to remain within the top right quadrant.

By focusing on core values we must get from a to b without exposing the bleeding edge technological advantage we possess. We can c further than our competitors by jumping straight to third base. We should not care if our strategy doesn't completely add up as long as we can square it with our key stakeholders. It's a win win!

Inscribed square proof

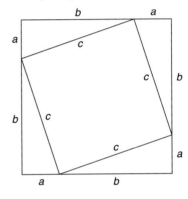

The figure shows a square of side $a + b$, so having area $(a + b)^2$. The divisions show that this is made up of a square of side c and four right-angled triangles of base a and height b. Hence the total area is also $c^2 + 2ab$, as each of the triangles has area $\frac{1}{2}ab$. Hence, since $(a + b)^2 = c^2 + 2ab$, simplification gives $a^2 + b^2 = c^2$.

Albert Einstein

$$E = ma^2 + mb^2 = mc^2.$$

Percy Bysshe Shelley

I set a triangle in an antique land
and said: Two vast and right-angled sides of stone
Stand in the desert. Near them, on the diagonal,
Half sunk, a vast hypotenuse lies, whose square,
Tells well it is those other two sides combined
When squared and summed as their sculptor well
    read

Truth which yet survives, stamped on these lifeless
  things,
A theorem that made them always so it's said.
And on the pedestal these words appear –
'My name is Pythagoras, king of maths:
Look on my works, ye Mighty, and be square'.
Nothing beside remains. Round the perimeter
of that forsaken shape, boundless and bare
The lone and level sands stretch in the
  perpendicular.

Station announcer
We regret to announce that triangle line services are
not running today. To get to c^2 customers are kindly
requested to take squares of services on the a and b
lines. Please take care to change at the right angle.
We are sorry for any inconvenience.

Daily Telegraph
Hypotenuse mystery solved: No Britons involved.

Agatha Christie
The little Belgian detective entered the drawing
room to greet the guests who had all mustered as
instructed, his thin black moustache freshly waxed,
his evening attire even more pristine than usual.
'Mes amis,' he began, 'I thank you for coming to
my little soirée.' 'Steady on old man,' interrupted the
Colonel, 'you had better explain yourself.' 'Indeed I
shall, indeed I shall. But first a little demonstration.
Monsieur Hastings, would you be so kind as to stand
by the window? And Mademoiselle Dupont, perhaps
you could stand in the corner by the fire place, ne
c'est pas.' They both did as instructed, although in
the latter's case not before she had had the opportunity to squeeze her fiancé's arm. 'And now Colonel,

if I could ask you sit at the writing desk where the
body was discovered'. 'Get on with it man', he objec-
ted, but grudgingly acquiesced. 'Patience, patience,
and all will be revealed. Et voilà, you see that you are
standing in a perfect right-angled triangle. From this
I think I can prove to you beyond any reasonable
doubt …' the detective paused, closed his eyes and
began twirling his moustache. 'If I could now call on
Stevens the butler to help me; if you would be so kind
sir, please pace out the distances between the Colonel, mon ami Hastings and the Mademoiselle and
tell me what you discover.' The butler did it.

Anagram
GO HEAR MATHS POETRY

Geometric proof

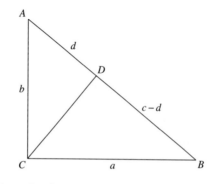

From the figure we see that the triangles ABC,
ACD, and DCB are similar and hence $\frac{d}{b} = \frac{b}{c}$, $\frac{a}{c} = (c-d)/a$. From the first, $cd = b^2$ and from the
second, $cd = c^2 - a^2$. Putting these together, $b^2 = c^2 - a^2$ or $c^2 = a^2 + b^2$.

CHAPTER 17

A conversation with Freeman Dyson

MARIANNE FREIBERGER AND RACHEL THOMAS

In February 2013 *Plus Magazine* was lucky enough to interview Freeman Dyson at the Institute for Advanced Studies (IAS) in Princeton, USA. Dyson (Fig. 17.1) is now 90 and still does physics every day in his first-floor office at the Institute. He told us the Institute will let him use the office as long as he can make it up the stairs! Dyson is a legend in theoretical physics. Among many other things he played a vital role in the development of quantum electrodynamics, an example of a quantum field theory designed to modify the classical theory of electromagnetism in the light of new insights from quantum mechanics. Initial versions of the theory were beset with mathematical difficulties: not only were calculations intractably complicated, they also tended to produce nonsensical infinities as answers! Dyson's mathematical rigour, together with ingenious and now famous diagrams produced by Richard Feynman, went a long way to solving these problems. Here is an edited version of our interview that we hope conveys his generous nature, wit, and intellect.

Plus: You started off in mathematics; how did you end up in theoretical physics?

Dyson: The fact was, I was never really a real mathematician. I was always an applied mathematician that happened to be in number theory, which is of course supposed to be pure mathematics. Number theory is really applied mathematics [using techniques], mostly from the 19th century, to find out things about numbers. It's similar to what you do in physics, [using mathematics] to find out things about atoms. So the switch to atomic physics really wasn't a big switch. It was essentially the same mathematics being used, just on different problems.

I was always interested in physics, but I came to Cambridge as a student in 1941 in the middle of the war. All the applied [mathematicians] were away fighting the war so there was essentially no physics going on. But there were these very famous pure mathematicians – such as G. H. Hardy and John Littlewood – whom I had all to myself. It was a wonderful time to be a student with these big, famous mathematicians around. They had hardly any students so I got to know them all and became, for the time being, a pure mathematician.

But I hadn't lost the interest in physics and after the war I made the switch. I found Nicholas Kemmer in Cambridge, who gave me this book [his treasured 70-year-old copy of

Fig 17.1 Freeman John Dyson (born 1923). Photograph courtesy of Freeman Dyson.

Gregor Wentzel's *Quantum theory of fields* (in the original German), that he'd retrieved from a nearby shelf at the beginning of our interview, referring to it as his 'bible']. It was Wentzel's book, published in Vienna during the war. It was a precious treasure, that book; I think there were only two copies in England at that time and I had one of them. Kemmer was supposed to be my colleague, but he knew all about quantum field theory [the mathematical language used to describe quantum physics], because he had been a student of Wolfgang Pauli before the war. Kemmer was a tremendous piece of luck for me, he was probably the best-informed person there was in England.

America was very empirical, nobody had heard of this stuff. They considered it a sort of Italian opera – extravagant and irrational entertainment – they didn't consider it useful. So I came to America as a student in '48 and I was then the teacher, teaching all these famous people, Hans Bethe and Richard Feynman, about quantum field theory.

The years of 1947–48 were a great time. It was just enormous luck to come into physics at that moment, because the experiments were roaring ahead. In those days it didn't take ten years to do an experiment, it took about ten days.

Microwaves caused a tremendous revolution in atomic physics. Instead of using just optical light to measure the [atomic] levels, you could do it a thousand times more accurately with microwaves.

For the first time [scientists] got a really accurate picture of the hydrogen atom, the supposed simplest atom. What they found disagreed with Paul Dirac's prediction. [Willis Lamb experimentally observed that two energy levels, which Dirac predicted would be the same, were in fact different.] This is the famous Lamb shift – which was discovered just when I came to America.

Everybody was very excited about [the Lamb shift]; for the first time there was a real discrepancy with the theory [quantum physics]. Bethe solved the problem as far as the physics was concerned. But the problem was if you calculated [Bethe's explanation] with the existing [mathematics], it [gave you infinite answers], and that was the big question.

All the giants from the old times were still around, people like Heisenberg, Schrödinger, Dirac, and Oppenheimer. They all thought we needed radically new physics [to solve this problem], they all had their theories of changing the whole basis of physics, introducing completely new ideas. But every one of them turned out to be wrong.

There came along these three bright young men, Richard Feynman and Julian Schwinger and Sin-Itiro Tomonaga, and each of them solved the problem [of the infinities] in his own way. All of them got the same answer, which obviously was right. I had the tools of quantum field theory, which they didn't have. So I was able to put them all together and demonstrate it was all quite simple. I polished up the mathematics so that it did give finite answers.

Plus: So do you think it was your mathematical background that helped?

Dyson: Yes. Schwinger of course was good at calculations, but he didn't like quantum field theory. He was this young prodigy, and he came and gave this talk here [at the IAS], explaining his calculation, and Oppenheimer said 'You know when other people give talks it's to tell you how to do it. But when Julian Schwinger gives a talk, it's to tell you that only he can do it.'

Feynman had his completely original way of doing things, which he didn't think of as quantum field theory, but it really is. Usually the way physics is done, since Newton, is you write down equations, the laws of physics, then you calculate the results. Feynman skipped all that, he just wrote down the pictures [known as Feynman diagrams] and then wrote down the answers. There were no equations. These Feynman diagrams were his invention and turned out to be in fact a pictorial description of quantum fields.

Plus: What was Feynman like as a person?

Dyson: A great guy. The joy of Feynman was that he was totally outspoken. He always said exactly what he thought about you or about anything else. So if I wanted to go and talk with Feynman, I could walk into the room and he would say 'Get out, don't you see I'm busy.' So I'd know that it wasn't a good moment and that was fine.

Another time I'd come in and he would start talking, be very friendly, so I'd know that he really welcomed me. It was an easy relationship, but he could be very brutal. But that, I found, made life simple. I always knew where I stood, and very often he'd like to go for a walk and asked me to come along. So we'd go for long walks and he would talk about all sorts of things. I enjoyed him very much, because he was a real performer. He just loved to perform and he had to have an audience.

Plus: Where did Feynman's concept come from? Where do revolutionary ideas come from?

Dyson: It's true of almost every great idea that you really don't know afterwards where it came from. Our brains are random, that's of course nature's trick for being creative. I have identical twin grandsons, they have all the same genes but they don't have the same brains: they develop independently. So these two identical young men have totally different brains, all the internal structure is essentially random. And that's how our minds turn out to be so powerful: they don't have to be programmed, they can invent things just by random chance. I think that's where [great ideas] come from. All really good ideas are accidental. There's some random arrangement of things buzzing around in somebody's head, and it suddenly clicks.

Source

An earlier version of this article appeared in *Plus Magazine* (<http://plus.maths.org>) in June 2013.

CHAPTER 18

A glass of bubbly

PAUL GLENDINNING

T he next time you are presented with a glass of champagne, look at the bubbles. If you don't have immediate access to a glass, look at the photograph in Fig. 18.1, which shows a stream of bubbles rising in a glass of champagne (Saumur to be honest, but *méthode traditionnelle*). The bubbles grow as they rise and the distance between bubbles increases as they move up the glass. Assuming that the rate of nucleation of the bubbles is constant, this means that the bubbles also accelerate. Teasing out why this happens is a nice exercise in mathematical modelling, although it could be argued that these observations and questions differentiate the scientist from the non-scientist; the latter is quite possibly more interested in either drinking the liquid or discussing whether the glass is half full or half empty.

I learnt about this subject from a talk by Richard Zare, a chemist from Stanford University. He mentioned in passing work he did some time ago on the growth of bubbles as they move up a liquid such as beer or champagne.

I am not unfamiliar with beer (although less at home with champagne), but to the extent that I had given the matter any thought I would have said that the increase in size of the bubbles as they rise is due to the reduction in pressure they experience, the pressure being proportional to the depth of the bubble. Not so, Zare informed the audience. The pressure effect is small compared to the influx into the bubble of further dissolved carbon dioxide from the liquid as the bubble rises. Moreover, Zare has also explained why the bubbles in a glass of Guinness (or other dark, heavy beers) appear to descend, and the same set of thoughts made me wonder about the trains of bubbles observed in less dense beers and more fizzy wines.

It turns out that the study of bubbles in champagne has received a lot of attention over the past few years, particularly from the group of Gérard Liger-Belair at the University of Reims, with support from Moët et Chandon and Champagne Pommery. Simply put, carbon dioxide (CO_2) is dissolved in the champagne under pressure. After pouring, gas bubbles form, or *nucleate*, on particles or fibres clinging to the surface of the glass. These bubbles then grow by diffusion of CO_2 molecules through their surface. Eventually the buoyancy forces are large enough for the bubbles to detach from the glass and

Fig 18.1 Bubbles in a glass of champagne. (Photograph by the author.)

the nucleation process can start again. We shall ignore nucleation here and concentrate only on the rising stream of bubbles. There are two things that we need to explain: why do bubbles grow and why do they rise? We shall use some elementary physics, which can be translated into mathematical equations.

Taking the first problem, once a bubble has detached from the glass it experiences buoyancy and drag forces. It also changes as newly diffused CO_2 enters it. This latter effect leads to the increase in size of the bubble as it rises. The rate of increase of CO_2 molecules in the bubble is proportional to its surface area $4\pi r^2$, so if the radius of the bubble is r – which is a function of time – then the number N of molecules satisfies an equation of the form

$$\frac{dN}{dt} = \gamma 4\pi r^2,$$

where γ is approximately constant and is proportional to the CO_2 concentration difference which drives the diffusion across the surface. Here the *differential* on the left-hand side of the equation means the *rate of change* of the number of molecules.

Assuming that the carbon dioxide in the bubble is an ideal gas (a gas made up of non-interacting particles which move randomly), its pressure p, volume V, and temperature T are related to N by the classic *ideal gas law* $pV = kNT$, where k is the *Boltzmann constant*, $k = 1.3806503 \times 10^{-23}$, measured in units of kilogram metres squared per kelvin second squared.

We can rewrite this equation as $N = pV/kT$. Assuming that the temperature T is constant and that changes in pressure over time are small compared to changes in volume over time gives a second equation for dN/dt (try it!), which when equated with the first equation implies

$$\frac{dN}{dt} = \gamma 4\pi r^2 \approx \frac{p}{kT} \frac{dV}{dt}.$$

We can now solve this equation using high school calculus. Since the volume of a sphere is $V = 4\pi r^3/3$, the *chain rule* for differentiation implies that $dV/dt = 4\pi r^2(dr/dt)$. Substituting this into the above equation and rearranging gives us

$$\frac{dr}{dt} \approx \frac{\gamma kT}{p} = c \quad \text{(constant)}.$$

This tells us how our bubble grows over time: the rate of change of r is approximately constant (the right-hand side of the last equation) and so, after time t, $r \approx r_0 + ct$, where r_0 is the initial radius of the bubble after nucleation.

So much for the growth of the bubble, but what about its upward motion? This can be analysed using the classic Newtonian law, 'force equals mass times acceleration', adapted to the fluid context, with two extra elements: *perturbation theory* (an applied mathematics technique to deal with quantities of different sizes) and *empirical approximations*, used by engineers to describe features that are not fully understood. This makes it typical of many problems of mathematical modelling where an absolute 'perfect' answer is beyond our current understanding (or simply too complicated to be useful).

The bubble experiences two forces, a buoyancy force upwards and a drag force against the direction of motion. Deriving an expression for the buoyancy force is straightforward. By

Archimedes' principle the bubble experiences a force $F_b = \frac{4}{3}\pi r^3 \rho g$ upwards due to the liquid it displaces, where g is the acceleration due to gravity and ρ the density of the liquid. The bubble experiences a similar force downwards as the density of the liquid is replaced by the density of the bubble, but this is negligible.

The drag force F_d experienced by the bubble is rather harder to derive. To a first approximation the frictional force goes like the square of the speed of the bubble and is proportional to the effective surface area presented, so

$$F_d(r, v) \approx -\frac{1}{2} C_d \rho \pi r^2 v^2.$$

The minus sign indicates that the force acts downwards, opposing the motion, and the term C_d is the *drag coefficient*, to which we will return in a moment.

In principle then, the sum of the forces acting on the bubble equals its 'mass times acceleration', but in fact this latter term is small compared to either of the forces and so it can be ignored to a first approximation (mathematically justified using perturbation theory), and so the buoyancy force and the drag force cancel approximately, so

$$C_d v^2 \approx \frac{8}{3} g r.$$

The big unknown in this equation is the drag coefficient C_d. It depends on the exact properties of the object that is moving through the liquid, such as its size and shape, as well as on the properties of the liquid. There are various alternative approximate expressions for the drag coefficient for a fluid sphere, such as our bubble, which have been determined experimentally. One that is particularly convenient for our example is $C_d = A(rv)^{-3/4}$, where the 'constant' A depends on properties of the fluid such as its viscosity and density. Substituting this expression for C_d into our earlier formula and rearranging to make the speed v the subject gives

$$v \approx \left(\frac{8g}{3A} \right)^{4/5} r^{7/5}.$$

If z is the height of the bubble then the velocity v is the rate of change of height, $v = dz/dt$.

We can again use high school calculus to combine this velocity, the rate of change of height with time dz/dt, with the expression dr/dt for how the bubble's radius r grows with time that we derived earlier to get an expression for the quantity we really need, dz/dr, which describes how a bubble's height changes with its radius. This is another application of the chain rule. The resulting equation can be solved exactly to obtain

$$z \approx \frac{5}{12c} \left(\frac{8g}{3A} \right)^{4/5} \left(r^{12/5} - r_0^{12/5} \right),$$

where c is the constant rate of growth of the radius of the bubble. (You can have a go at deriving the equation and integrating it yourself.)

These expressions suggest that both the speed and the height depend on simple *powers* of the bubble radius (or, equivalently, since they are proportional, the time from nucleation), multiplied by constants that depend on the liquid's properties. Additionally, and rather pleasingly, there are experiments that seem to confirm these relationships.

Bubble problems are much harder and richer than I had imagined. This is a rough and ready analysis, based on experimental observations and so-called 'back of the envelope' calculations, which does nevertheless seem to apply nicely to champagne.

And what of the Guinness? According to Zare, the bubbles *are* falling. This is because in a glass which has sides that angle outwards there is a strong rising column of bubbles in the middle of the glass, and the displaced liquid falls back towards the sides of the glass, in a sort of convection pattern. This in turn pushes the bubbles down, overcoming their buoyancy forces. Since the liquid is dark, we only see these falling bubbles on the outside and not the much stronger rising columns in the middle. Andrew Fowler and his colleagues, mathematicians from the University of Limerick in Ireland, have studied this phenomenon in greater detail, presumably requiring much first-hand practical demonstration!

• •

FURTHER READING

[1] Gérard Liger-Belair and Philippe Jeandet (2002). Effervescence in a glass of champagne: A bubble story. *Europhysics News*, vol. 33, pp. 10–14.
[2] Marguerite Robinson, Andrew Fowler, Andrew Alexander, and Stephen O'Brien (2008). Waves in Guinness. *Physics of Fluids*, vol. 20, 067101.
[3] Richard Zare (2005). Strange fizzical attraction. *Journal of Chemical Education*, vol. 82, pp. 673–675.

Source

A longer version of this article appeared in the regular 'View from the Pennines' column of *Mathematics Today* in April 2006.

CHAPTER 19

The influenza virus: It's all in the packaging

JULIA GOG

Influenza is a seasonal disease we are all familiar with. Most of us have experienced the misery of being cooped up in bed with a streaming nose, aching joints, and temperature. However, sometimes influenza can cause serious complications, particularly in the elderly, occasionally leading to death. What makes influenza so difficult to control is the way in which it evolves. All viruses evolve, with minor genetic mutations either being carried forward to new generations or dying out. This normal form of evolution in the influenza viruses which circulate in humans is called *antigenic drift*: the virus slowly changes over time, and we retain some partial immunity to the evolved virus thanks to our immune systems remembering similar past infections. However, things become very dangerous when *antigenic shift* occurs. In this case two different strains of influenza mix together to create what is essentially a new disease to which we have no immunity.

During an influenza infection, viruses spread from cell to cell, hijacking cellular machinery to make more copies of themselves. The influenza genome is unusual in being made of several segments, rather than one continuous genome. The different segments are replicated separately, but must end up together inside a new virus particle – a process known as *packaging*. There are eight segments in the most common form of influenza, and all contain important genes, so a virus needs a full set of segments to be fully viable.

Being segmented, it is easy for influenza to swap genes between different strains: a process called *reassortment*. Reassortment is the mechanism behind antigenic shift and was implicated in all the 20th-century flu pandemics. The 2009 swine flu virus is believed to be a multiple reassortant, where several past reassortment events led to the virus that could spread across the globe.

Let me count the ways

There was debate in the past about whether influenza packages segments at random or whether there is a special mechanism to ensure each virus particle gets the eight different segments. But the numbers immediately suggest that random packaging would be a truly pathetic strategy for a virus! If we number the segments 1 to 8, then random packaging is equivalent to picking eight numbers at random from this set, say {5, 5, 7, 4, 1, 6, 3, 8}. But this example won't do, as it doesn't contain all eight segments. There are $8^8 = 16,777,216$ ways of picking a random set of eight numbers, with all of these random sets equally likely to be picked. But the number of ways of picking exactly one of each of the segments is

$$8 \times 7 \times 6 \times 5 \times 4 \times 3 \times 2 \times 1 = 8! = 40,320.$$

So the chance of picking a complete set at random is $8!/8^8 = 0.0024$, or less than 1 in 400. Random packaging clearly isn't a good strategy.

Looking for signals

Current experimental evidence suggests that segments are indeed packaged specifically: the virus has some way to ensure that it usually has exactly one of each segment. We don't yet know exactly how this happens, but for packaging to be specific, there must be something about each of the eight segments that acts to distinguish it from the others: a label or *packaging signal*.

The influenza virus genome is made up of eight segments of RNA which vary from 890 to 2341 nucleotides in length. Nucleotides – adenine (A), uracil (U), guanine (G), and cytosine (C) – are the molecules that make up the strand of RNA. The RNA strand is read as a sequence of nucleotide triples, called *codons*: each codon defines an amino acid, and the sequence of amino acids the RNA specifies builds up a particular protein molecule.

Genetic code has redundancy: different codons can represent the same amino acid. Different influenza virus strains have different RNA sequences, and there is variation in which codon is used when there are possible synonyms available. However, if a packaging signal is embedded, it is like a message hidden within another message. Then, the choice of codon is not so free, so packaging signals should show up as regions where there is low variability.

To find the suspicious regions of the genome segments, we align hundreds of examples of the genome sequence (which are publicly available). For example, the following sequences are all from one segment of the influenza virus:

```
a  u  g  g  a  G  a  g  a  a  u  a  a  a  A  g  a  a  c  u  a  A  g  A  ...
a  u  g  g  a  A  a  g  a  a  u  a  a  a  A  g  a  a  c  u  a  C  g  G  ...
a  u  g  g  a  A  a  g  a  a  u  a  a  a  A  g  a  a  c  u  a  C  g  G  ...
a  u  g  g  a  A  a  g  a  a  u  a  a  a  A  g  a  a  c  u  a  A  g  A  ...
a  u  g  g  a  G  a  g  a  a  u  a  a  a  G  g  a  a  c  u  a  A  g  A  ...
```

These break down into codons as follows:

```
aug   gaG   aga   aua   aaA   gaa   cua   AgA   ...
aug   gaA   aga   aua   aaA   gaa   cua   CgG   ...
aug   gaA   aga   aua   aaA   gaa   cua   CgG   ...
aug   gaA   aga   aua   aaA   gaa   cua   AgA   ...
aug   gaG   aga   aua   aaG   gaa   cua   AgA   ...
```

Despite slight differences in the sequences (in the capitalised positions), all these sequences encode the same series of amino acids:

```
Met   Glu   Arg   Ile   Lys   Glu   Leu   Arg   ...
```

All the sequences must start with methionine (Met), encoded **aug**, as this codon is used to indicate the start of the reading frame. This codon can't tell us anything about packaging signals.

The next codon is glutamic acid (Glu), which can be encoded as either **gag** or **gaa**. As both of these variations are present in the genome sequence, this codon is unlikely to be part of the packaging signal.

The third codon is arginine (Arg). There are six possible ways of encoding arginine: **cgt**, **cgc**, **cga**, **cgg**, **agg**, as well as **aga**. However, as there is no variation in this codon for these sequences, we would suspect this region might act as a packaging signal.

In fact, we can come up with a score for every position in the sequence (based on the number of variations possible, as well as some other factors, such as codon bias), where low scores mean there's suspiciously little variation when variation would have been possible:

$$
\begin{array}{cccc}
aug & gag & aga & \dots \\
aug & gaa & aga & \dots \\
1.00 & 1.00 & 0.01 & \dots
\end{array}
$$

Why do we care?

Once we have pin-pointed suspicious regions in the genome our colleagues at the Department of Pathology, University of Cambridge, explore them further in the lab by engineering mutant viruses to see if the packaging process is broken by changes in those locations in the genome. It seems that our methods are identifying regions of interest. Our other work has included applying our approach to explore rotavirus (which causes diarrhoeal disease and is a major cause of death in young children worldwide), which also has a segmented genome. Current research is to extend our methods to look for interesting regions in the HIV genome, in collaboration with the Department of Medicine at the University of Cambridge.

Understanding the packaging process of influenza and uncovering the packaging signals that drive it would be a major step in understanding how viruses work. Not only is this of basic virological interest, but it could also lead to possible treatments. And, importantly, it would lead to a better understanding of the reassortment of 'normal' flu with avian or swine flu that allows dangerous viruses to become adept at infecting humans. Perhaps this research will give us another defence against pandemics in the future.

Source

This article originally appeared in *Plus Magazine* (<http://plus.maths.org>) on 8 December 2009.

CHAPTER 20

Solving the Bristol bridge problem

THILO GROSS

It's 6.38 a.m. on 23 February 2013. At this time on a Saturday there is hardly any traffic. The silence is broken only by the calls of the seagulls and the sound of my steps as I climb the old iron stairs leading up to Vauxhall Bridge. The riveted footbridge takes me across the waters of the New Cut to a part of Bristol called Bedminster. Once there, I turn left and walk upstream until I reach another footbridge, Gaol Ferry Bridge. Compared to the steam age aesthetics of Vauxhall Bridge, this bridge seems impossibly light and playful with its lavish Victorian decorations. I cross the bridge back to Spike Island, in the centre of Bristol. The next bridge is now only a few steps away, a swing bridge, imaginatively called Swing Bridge.

My walk is inspired by a famous mathematical problem, solved in the year 1736, that relates to a city then known by its German name, Königsberg. Captured by the Soviet Union during the Second World War and renamed Kaliningrad, the modern-day city is completely surrounded by Poland and Lithuania and forms a small Russian enclave on the Baltic coast. The old city to which the problem refers lies close to the mouth of the river Pregel, occupying both banks of the river and two river islands. Connecting the different city districts were seven great bridges, and a much-debated question was whether it was possible to go for a continuous walk that took the walker across each bridge exactly once. Many citizens claimed that they had found such a walk, but when challenged could not reproduce it without crossing one of the bridges twice. Others held that such a walk did not exist, but could not find a way to prove it either. By 1736 the bridge problem was not only discussed in Königsberg but also occupied the minds of the mathematical elite all over Europe.

I have decided to take on the same challenge in my new home, the city of Bristol, which has many similarities to Königsberg. It is of more or less the same population as modern Kaliningrad, and has a similar maritime history. Bristol lies close to the mouth of a river, the Avon, and, like the old city of Königsberg, it occupies both banks of the river and two river islands (see Fig. 20.1).

Meanwhile Swing Bridge has taken me to Redcliffe, on the second island. I follow the New Cut further upstream until I reach Bedminster Bridge. This old road bridge has been extended into a roundabout by the addition of a second bridge side-by-side to the original. I use the older bridge to cross the New Cut to Bedminster and return to Redcliffe via the newer bridge. Further up the New Cut I pass Langton Street Bridge, whose yellow colour and bent shape have earned it the

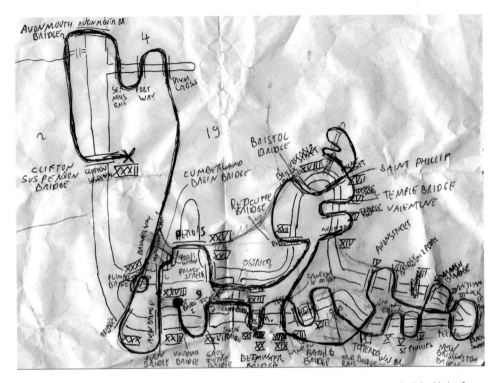

Fig 20.1 Euler's solution of the bridge problem, crossing every bridge in a city exactly once, marked the birth of graph theory. The *Eulerian* walk that solves the problem is a global object that depends intricately on every bridge, whereas a failure to complete the walk occurs because of local problems that are easier to analyse. This allowed Euler to formulate precise conditions for the existence of the walk. However, finding an actual solution that is walkable with reasonable effort still requires a bit of tinkering. Shown in this figure is a map that has evolved through several versions. After several scouting tours and one failed attempt, this map accompanied me on a walk that crossed every bridge in the city of Bristol exactly once.

local name of the Banana Bridge. But I can't cross this bridge now, because it will be needed later. Instead, I continue upstream to Bath Bridge, another double-bridge roundabout, where I cross over to Bedminster and back to Redcliffe once more.

To be honest, I will not cross every single bridge in Bristol. There are some bridges that cannot be legally crossed on foot and there are numerous small bridges that cross ditches and streams. But I figure that similar minor bridges must surely have existed in Königsberg and were not included in the original problem, which was all about a leisurely walk for the lords and ladies around the heart of the city. It is thus clear how to define the Bristol problem. My challenge is to cross precisely once every bridge that spans one of Bristol's major waterways and is legally walkable. Unfortunately that still leaves me 42 bridges to deal with. It's going to be a long day.

Continuing upstream I reach St Philip's Marsh, which was once a separate island, but is now connected to Redcliffe. At this point also the artificially dug river bed that is the New Cut ends and turns into the Avon proper. A little further lies an old railway bridge. This bridge derailed one of my earlier attempts at the bridge walk, when I found that it was crossable although I had thought it was not. Finding this bridge meant that I had started the walk in the wrong place and could not complete it without crossing a bridge twice. However, this time I have carefully planned my route beforehand and nothing is going to stand in my way. I use the railway bridge to cross the Avon and continue upstream. I pass Totterdown Bridge, another one that will be needed later, and then cross the Avon again on Greenway Bridge, a beautiful wooden footbridge.

What makes the bridge problem fascinating is that it contains a taste of the branch of modern mathematics that is sometimes called complex systems theory. This is the theory of how order emerges out of hierarchies of masses of interacting objects. Although the bridge challenge can be simply phrased, the walk that solves the challenge is a complex object that depends intricately on the placement of every single bridge. One can say that the walk emerges from the pattern of bridges. Dealing mathematically with such emergent properties of interconnected systems is the core challenge that underlies a wide variety of phenomena, ranging from tackling climate change to understanding opinion formation in social networks.

Meanwhile I have reached the St Philip's Causeway Bridge, another double bridge, on which I cross the Avon two more times. Then I turn my back on the Avon and cross the breadth of St Philip's Marsh to the Feeder Canal, which connects the Avon to Bristol's Floating Harbour. I cross the Feeder Canal on Marsh Bridge and walk further upstream to the Netham Lock Bridge, the smallest double bridge on the walk. I must cross only one of the two bridges right now. This takes me back across the Feeder Canal to the very tip of St Philip's Marsh. From here it is just a few steps back to the Avon, which I cross on New Brislington Bridge, before heading yet further upstream. Three and a half hours into the walk, I reach St Anne's Bridge, the eastern-most point of the walk. From here on there are no further bridges across the Avon until the town of Keynsham. I turn downstream, and follow the Avon to the beginning of the Feeder Canal. Walking down the canal I cross the second bridge at Netham Lock, the Barton Hill footbridge, and another bridge named Marsh Bridge. Bristol's newest bridge is now just around the corner, but I can't reach it because I am on the wrong side of the Feeder Canal. It is time for a detour across some bridges which we have already seen but not yet crossed. I walk back across the breadth of Saint Philip's Marsh and cross the Avon on Totterdown Bridge. Back on the Bedminster side it is a short walk through parks and residential areas to Langton Street's Banana Bridge.

In principle, the bridge problem can be solved by a brute force approach. We know that the solution that we seek must be a sequence of bridges containing every bridge exactly once. We can thus list all such sequences and then, for each sequence, check whether it is actually walkable. In Bristol one such sequence could start 'Swing Bridge, St Anne's Bridge, Vauxhall Bridge, ...'. But this sequence is not walkable, because Swing Bridge connects Spike Island and Redcliffe, whereas St Anne's Bridge is not reachable from either of the two islands without crossing another bridge. Checking all sequences in this way, we either eventually find the desired walk or prove that the walk does not exist.

To solve the Königsberg problem, we only have to check 5040 sequences. To see this, consider that when choosing the first bridge in the sequence we have a choice among seven bridges; once the first bridge in a sequence is chosen, we have six choices left for the second bridge and then five for the third and so on. Thus the total number is $7 \times 6 \times 5 \times 4 \times 3 \times 2 \times 1$, which we can write as 7! (seven factorial). Such brute force solutions have become very popular because they can be quickly implemented on computers, which can run through the 5040 sequences in some microseconds. However, even in 1736, we could have sat down with quill and parchment to write out and check all the sequences within a few days. For Königsberg, this would have revealed that the desired walk does not exist. We could still have gone off to our peers and claimed that we had resolved the problem by proving this once and for all. However, this would have hardly earned much bragging rights, because said peers would have to go through the whole list of 5040 sequences again to confirm our results. For Bristol, we need another solution anyway. For the 42 bridges of Bristol, one would have to check 42! sequences – a number with 51 digits. Even the fastest computers would need more than a million trillion trillion times the age of the Universe to check that many sequences.

Back on the walk I have reached Bristol's newest bridge. The modern footbridge is made of polished steel and has many tiny holes that are lit at night. But now it is noon, and the bridge reminds me of a giant cheese grater. I use the bridge to cross Bristol's Floating Harbour, the name given to the Avon's former river bed, now kept isolated from the huge tidal range by lock gates and used mostly by pleasure craft. I immediately return to Redcliffe by crossing Valentine's Bridge, another modern footbridge. I wind my way further down the Floating Harbour, crossing the Temple Bridge and St Philip's Bridge and taking a small detour to take a little footbridge across the remainder of the moat of Bristol castle. Then I reach Bristol Bridge. The great stone arches of this bridge were completed in 1768, but most of the upper half of the bridge was added in Victorian times. The first stone bridge in this place was probably built in the 13th century. Its wooden predecessors gave the city its name *Brycgstow* – place of the bridge.

In 1736, 32 years before the completion of Bristol Bridge, the Königsberg bridge problem was solved by Leonhard Euler, the pre-eminent mathematician of the 18th century who introduced a vast array of what we now take for granted in mathematics and mechanics. The key insight in his solution is that the only way of failing to find an acceptable walk is to trap oneself in a city district that one cannot leave without using a bridge twice. While the walk that we seek is a complex global object, failure occurs because of a local problem, and hence is much easier to analyse.

Consider for instance a district that can only be reached via a single bridge. If we do not start in this district, our walk must end there, because once we cross its bridge we cannot leave again. So any district with only one bridge must be the start or end point of the walk. Neither Königsberg nor Bristol has such a district with only a single bridge. But the same conclusion holds for all districts that are served by an odd number of bridges. If a district has three bridges, we must visit it twice to cross all its bridges. However, if this district wasn't our starting point, there will only be a single usable bridge left after we have arrived and departed for the first time, and thus the walk will end on the second visit. And so on and so forth. A district with 13 bridges needs to be visited seven times, and if the district was not our starting point, the walk will end when we arrive for the seventh time. Thus, every district that is served by an odd number of bridges can only be the start or end point of the walk. Since the walk can only have one start and one end point, a walk cannot exist if more than two districts have an odd number of bridges.

Meanwhile I have used Redcliffe Bridge to return to Redcliffe a final time, before taking the blue Ostrich Bridge to Spike Island. I cross the Floating Harbour again on Prince Street Bridge and then follow it downstream while crossing Pero's Bridge and Poole's Wharf Bridge. Cumberland Basin Bridge takes me across the Floating Harbour once more, back to Spike Island. Crossing the breadth of the island I reach Avon Bridge, a small railway bridge, designed by Isambard Kingdom Brunel. After crossing this bridge it is just a minute's walk to Brunel Way, one of the major roads into Bristol. Brunel Way crosses the Avon and the Floating Harbour on a giant swing bridge, called Plimsoll Bridge. The first arc of this bridge takes me back to Spike Island, where I take a small detour to cross two smaller bridges beneath Plimsoll Bridge. Then, via the second arc, I cross the Floating Harbour once more and arrive in the district of Hotwells. Unfortunately, my next bridge is now five miles away.

In Königsberg, all four districts are served by an odd number of bridges and thus Euler's solution proved that no walk can exist. In Bristol, Spike Island and the right riverbank have an odd number of bridges, whereas the left bank and Redcliffe are served by an even number of bridges. Thus a walk exists and must start on Spike Island or the right bank. I have chosen the right bank as the end point of the walk, as this is served by Bristol's most famous bridge and major landmark, Brunel's Clifton Suspension Bridge. Ending the bridge walk with the Suspension Bridge has not been possible until last year, when Bristol's youngest bridge was built. A new footbridge, named Möbius Bridge, is scheduled to be built soon, which will make ending the walk with the Suspension Bridge impossible once more.

A criticism often levelled at mathematicians is that they occupy their time by thinking about irrelevant simple puzzles. Indeed, it can be argued that Euler's elegant solution of the bridge problem has little practical applicability. Nevertheless, the wider importance of his work is hard to overstate. The solution of the bridge problem marked the birth of graph theory, the mathematical study of networks, and by extension the birth of discrete mathematics and the modern subject of network science. It thus led to a series of discoveries that provide the basis for Google, Facebook, Twitter, and even the Internet itself. Perhaps even more importantly, it is becoming apparent that networks provide a unifying theme for research into complex systems and thus have huge importance, for example in biology, medicine, and the design of robust technological systems.

Meanwhile I have reached Sea Mills, where I cross the River Trym three times. From here it is another four miles' walk along the bank of the Avon to the largest bridge in the walk. The Avonmouth Bridge carries the M5 motorway across the river. From atop the almost mile-long bridge both Bristol and Cardiff are visible, and in the distance the two huge motorway bridges that cross the River Severn can be seen. From here on I walk upstream again. Six more miles to the final bridge.

The sun is already setting when I reach Clifton Suspension Bridge. From the small footpath on the bank of the river the bridge appears as an impossibly thin band that crosses the Avon Gorge high above. I first pass under the bridge and then, after a quarter of a mile, double back onto a forest path that climbs the 200 feet up to the level of the bridge. It is already dark, and the bridge is beautifully lit by tiny white lamps that trace the curve of the steel-link cables. After 33 miles I cross the final bridge, tired but satisfied.

CHAPTER 21

All ravens are black: Puzzles and paradoxes in probability and statistics

DAVID HAND

You've made it through to the final of a TV game show and a million-pound prize beckons. Your host, named Monty Hall, shows you three closed doors. Behind one lies life-changing riches. Choose one of the other two and you win nothing. Underneath the studio lights, you feel the perspiration sliding down your face. It's only the background music, building to a crescendo that drowns out your beating heart. But then the music stops. Monty Hall turns to you and asks you to select your door. It's all you can do to get the words out, but you force them from your lips, choosing door number one. Monty Hall smiles. This being television, he's not about to put you out of your misery. Not just yet.

Instead, knowing what lies behind each of the three doors, Monty calls for door number two to be opened. There's nothing behind it. All along, the million pounds was behind either door one or door three. The genial host turns to you and asks if you want to stick with your original choice, or change your mind and open door three instead. With a choice of only two doors and the money behind one of them, you pull yourself together. Intuition tells you it's a 50:50 choice. If you change your mind now and get it wrong, you'll look an idiot in front of the watching millions, and besides, if the probability is the same that either door hides the money, there's no benefit in switching. You stick to your guns and stay with door one, confirming it's your final answer. Monty smiles, a little sadly this time. The door opens and you have lost. As the show's credits roll, security staff step forward to remove the million pounds behind door number three that could so easily have been yours.

Your intuition that there was a 50:50 chance that either of the remaining two doors hid the cash was not right. (If you don't believe me, try simulating it.) Your idea of the basic probabilities was rooted in a false assumption. In fact, your probability of choosing the correct door initially was 1/3, and the probability that one of the other two doors hid the money was 2/3. If you switch doors after the presenter knowingly eliminates one of the two losing doors, you win if your initial guess was wrong, which it was with probability 2/3. If you stick with your original choice, you win with probability 1/3. You should switch.

The *Monty Hall problem* has stimulated a great deal of debate – a whole book has been written about it, *The Monty Hall problem: The remarkable story of math's most contentious brainteaser.*

Even the great mathematician Paul Erdős got the answer wrong when he first encountered it. Much of the debate has arisen because there are different versions of the problem. For example, had the presenter picked his door at random, which he then opened and showed to conceal nothing, the solution would be different: in this case, the probability of your door and the unopened one containing the money are equal. The confusion arises from ambiguity about exactly what the question is, and this is also the source of the power of mathematics. Mathematics strips away the ambiguity, revealing the essence of problems, and of arguments.

If probability has its counter-intuitive moments, statistics seems to suffer from them even more, at least in terms of public image. The fact that Disraeli may have been contrasting 'lies and damned lies' with the truth revealed by 'statistics' does not seem to be a widespread understanding of the familiar adage: most people seem to think he was classifying statistics as a third, particularly egregious, kind of lie.

One example of a counter-intuitive statistical idea is known as *Simpson's paradox*. At its simplest level, this involves comparing the relationships within two groups with the relationship when the two groups are combined. For example in 1977 psychiatrists Donal Early and Michael Nicholas were exploring the demographic changes to the population of a psychiatric hospital over time:

Year	Males	All patients	Proportion of males
1970	343	739	0.464
1975	238	515	0.462

They found an apparent but very slight reduction in the proportion of male patients, from 0.464 in 1970 to 0.462 in 1975. To see whether this slight reduction was mainly down to a fall in the proportions amongst younger or older patients, we can split the population into two, examining how the proportions changed for those under 65 and those 65 or over:

Year	Age group	Males	All patients	Proportion of males
1970	Under 65	255	429	0.594
	65 and over	88	310	0.284
1975	Under 65	156	258	0.605
	65 and over	82	257	0.319

This is odd. We find that the proportion of males in *both* the younger and the older groups has increased (from 0.594 to 0.605 for the under-65s and from 0.284 to 0.319 for the over-65s). The answer to the question 'was it mainly the change in the younger or older age group that led to the decrease?' appears to be 'neither: both led to an increase'. How can this be?

Once again a careful examination of things reveals what is going on. The puzzle arises because we intuitively expect the proportions aggregated over both age groups to be the average of the proportions within each age group. But this is only half true. The overall proportions are indeed averages but they are weighted averages, where the weights are the proportion of the population in each age group. And the different age groups have different weights in 1970 and 1975.

The overall proportion of males is

(proportion of population which is younger) × (proportion of males within younger group) + (proportion of population which is older) × (proportion of males within older group).

For 1970 this becomes

$$343/739 = (429/739) \times (255/429) + (310/739) \times (88/310)$$

or

$$0.464 = 0.581 \times 0.594 + 0.419 \times 0.284.$$

In contrast, for 1975 it becomes

$$238/515 = (258/515) \times (156/258) + (257/515) \times (82/257)$$

or

$$0.462 = 0.501 \times 0.605 + 0.499 \times 0.319.$$

Although both age groups show a higher proportion of males for 1975 than 1970, the 1975 calculation balances the two age groups about equally in the overall average (with weights 0.501 and 0.499). In contrast, the 1970 calculation puts a much higher weight (0.581) on the younger group than the older (0.419). The overall average for 1970 is thus larger than that for 1975. Again, careful elucidation of the assumptions and the steps involved in reaching a conclusion explains the counter-intuitive paradox.

Our final, less familiar example is Hempel's *paradox of the ravens*. This begins with the truth that the statement 'all ravens are black' is logically equivalent to the statement 'all non-black objects are non-ravens'. If you need convincing that these are logically equivalent, draw a Venn diagram (Fig. 21.1). The first statement is reflected by the fact that the ravens bubble sits inside

Fig 21.1 Paradox of the ravens: Venn diagram.

the black-objects bubble. But that means that there are no ravens outside the black-objects bubble: all non-black objects are non-ravens.

Now it is generally agreed that observing a black raven is supportive evidence for the statement 'all ravens are black'. Again, if you need convincing, imagine looking at a thousand ravens, or a million, or a billion, and noting that they are all black. Surely, as the numbers stack up, one would be increasingly confident in the statement that all ravens are black. This means that each fresh observation of a black raven adds more supportive evidence.

Exactly the same form of argument also tells us that observation of a non-black non-raven is supportive of the statement 'all non-black objects are non-ravens'. However, that means that such an observation would also be supportive evidence for the statement 'all ravens are black' since the two statements are logically equivalent. But this does not seem right. It would mean, for example, that observing a red ball or a green pen would be evidence supporting the assertion that 'all ravens are black'. In general, it would mean that we could learn things about ravens by looking at balls and pens.

There have been many attempts to resolve this paradox, often involving deep philosophical speculations. But there is a very simple solution based on a little statistical understanding and, in particular, based on sampling theory. We consider two distinct approaches to observing the non-black non-raven. In the first, we pick a sample object from the set of non-ravens, and observe it to be non-black. This carries no information about the raven population: it is explicitly restricted to the non-raven population. In the second approach, we select a sample object from the set of non-black objects, and observe it to be a non-raven. This does give us information about ravens. Again, to see this imagine repeating the exercise many times: after a billion draws of non-black objects and observing each of them to be non-ravens we might begin to suspect that there were no ravens amongst the non-black objects, that all ravens were black.

So that is the power of mathematical and statistical arguments. They abstract the problem from the confusing ambiguity of the real world, focusing instead on an artificial world that aims to represent the pertinent features of the real-world problem. In doing so, they force us to make explicit our assumptions and the characteristics of the artificial world we have created. Then they allow us to draw correct conclusions about behaviour within the confines of our abstract world. And probability and statistics are only counter-intuitive if something is missing from our abstract model.

· ·

FURTHER READING

[1] D. F. Early and M. Nicholas (1977). Dissolution of the mental hospital: Fifteen years on. *British Journal of Psychiatry*, vol. 130, pp. 117–122.

[2] Jason Rosenhouse (2009). *The Monty Hall problem: The remarkable story of math's most contentious brainteaser*. Oxford University Press.

CHAPTER 22

The Tower of Hanoi: Where mathematics meets psychology

ANDREAS M. HINZ AND MARIANNE FREIBERGER

The Tower of Hanoi is an unassuming puzzle, you might think (see Fig. 22.1). However, beneath the surface lurk a wealth of beautiful mathematical features and some surprisingly tricky questions. And the game has another trick up its sleeve. By virtue of its simple rules and scope for variation, it's a tool popular with psychologists to assess people's cognitive abilities. This has led one of the authors of this article (AMH) on a somewhat unlikely excursion into the field of psychology.

But let's start with the maths. The rules of the game are as follows. There are three pegs and a number of discs, stacked up on one of the pegs in order of size, with the biggest disc at the bottom. Your task is to transfer the whole tower onto a different peg, disc by disc, but you are not allowed to place larger discs onto smaller ones.

The game plan

The best way to see the scope of the game is to draw a graph that displays all the possible configurations and moves. Suppose we play the game with three discs. Label the discs 1, 2, and 3, with

Fig 22.1 The Tower of Hanoi. Copyright 2001 H. Steinlein.

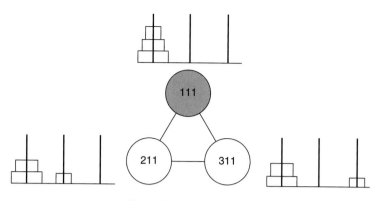

Fig 22.2 Connecting the dots.

1 being the smallest disc and 3 the largest one. Also, label the pegs 1, 2, and 3. Now suppose discs 1 and 2 are on peg 1 and disc 3 is on peg 2. You can encode this situation using the triple (1,1,2). The positions in the triple correspond to the discs and the value in a position tells you which peg the corresponding disc is sitting on. There's no confusion about the order in which discs sit on a given peg, because they have to be arranged in order of size. So all the legal configurations of discs can be encoded unambiguously in triples of numbers.

Now, for each triple, draw a dot on a piece of paper. Connect two dots if a single move can get you from one to the other. See Fig. 22.2 for an example. Altogether there are $3^3 = 27$ admissible positions. These can be arranged to give you the graph shown in Fig. 22.3.

This object is called a Hanoi graph and is denoted by H_3. The subscript 3 indicates that we are playing with three discs. Any path in the graph from $(1, 1, 1)$, which represents the discs in order on the first peg, to either $(2, 2, 2)$ or $(3, 3, 3)$, representing the discs in order on one of the other pegs, is a solution to the problem. The shortest such path is clearly one of the straight lines on the left-hand or right-hand boundary, each taking seven moves.

Adding discs

What can we say about the game with 4, 5, 6, or any number n of discs? In terms of the graphs, a very pretty picture emerges: the Hanoi graph H_4 for the four-disc version of the game consists of three copies of H_3, each connected to each of the other two by exactly one edge, in each case, as drawn in Fig. 22.4. This shows that the 'winning' strategy is to transfer three discs onto another peg using the H_3 strategy, then move the largest disc, and then transfer the remaining three discs onto the largest disc using the H_3 strategy again, making a total of $7 + 1 + 7 = 15$ moves.

Similarly, H_5 consists of three copies of H_4, H_6 consists of three copies of H_5, and so on: a beautiful nested structure emerges. This is due to the recursive nature of the game. If you ignore the biggest disc, the $(n + 1)$-disc version of the puzzle turns into the n-disc version. The biggest disc can sit on any one of the three pegs. The collection of possible moves for each of these three possibilities, using the remaining n discs, gives rise to one copy of H_n in each case. With a bit of thought, you can convince yourself that exactly one edge links any two of these copies.

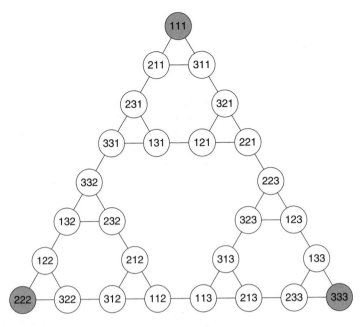

Fig 22.3 The Hanoi graph H_3.

This immediately poses an intriguing question. What can we say about the graph if the number of discs increases beyond any limit? Have a look at Fig. 22.5 (left).

This is Sierpiński's triangle, which you get from another infinite process. You start with a (filled-in) equilateral triangle and remove the middle (upside-down) triangle whose edges connect the midpoints of the three sides (you remove only the inside of that triangle, leaving behind its sides). You are left with three equilateral triangles and again remove the centre triangle from each of them, leaving you with nine triangles. Keep going, always removing centre triangles from what is left, ad infinitum. The object you get in the limit is Sierpiński's triangle.

Sierpiński's triangle is one of the most popular examples of a fractal. It is self-similar: if you zoom in on what is left of any given tiny triangle, what you see is exactly the same as the whole picture. The triangle also lives in a strange world 'in-between' dimensions: it is 'more' than a 1-dimensional line, yet its area is zero so it is not a 2-dimensional object either. In fact, its fractal dimension is $\log 3/\log 2 \approx 1.585$, which is between 1 and 2.

As you add discs to the Tower of Hanoi game, the corresponding graphs, suitably rescaled, start to look more and more like Sierpiński's triangle. And the object you get in the limit as n tends to infinity has exactly the same structure as Sierpiński's triangle!

Another famous triangle

There is an equally intriguing connection to another triangle beloved by mathematicians: Pascal's arithmetical triangle, which encodes the coefficients that appear in the expansion of $(x + y)^k$. If you take the first 2^n rows of Pascal's triangle and connect odd numbers that lie next to each other, either horizontally or diagonally, then the graph you get has exactly the same structure as the Hanoi graph H_n – see Fig. 22.5 (right).

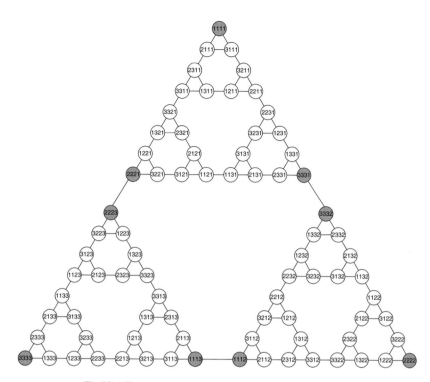

Fig 22.4 The Hanoi graph H_4 is composed of three copies of H_3.

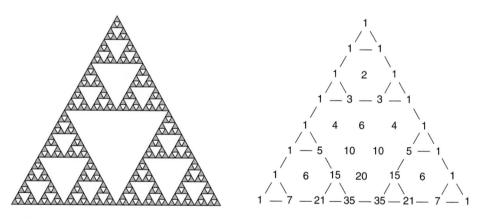

Fig 22.5 (Left) Sierpiński's triangle. (Right) The first eight rows of Pascal's triangle with adjacent odd entries connected.

These connections are not only beautiful, they are also useful. Results that are hard to prove for one of these objects may be easier to prove for another and can then be transferred. For example, consider the distance between any two points in Sierpiński's triangle. What would the average of these distances be? This question troubled mathematicians for some time, but using Hanoi graphs it is possible to discover the answer: it's $466/885 \approx 0.53$ (assuming that the side length of the initial triangle in the construction of Sierpiński's triangle is 1).

So much for adding discs, but what happens when you introduce another peg? The game becomes easier because you have more scope for moving discs around. But the graphs also lose their neat structure. There are now more configurations of discs that allow you to move the largest disc; the smaller ones no longer need to be stacked up on a single peg.

This makes the graphs more complex. For instance, with four pegs (and more than two discs) the graphs are not planar anymore: you cannot draw them on a sheet of paper without the edges crossing. Mathematicians still don't understand these graphs very well, because they are so strongly intertwined. This means that seemingly simple questions can be surprisingly hard to answer. For example, nobody knows how long the shortest solution is for the multi-peg puzzle. There are strategies for solving it, and the notorious Frame–Stewart conjecture says that these strategies are optimal. But although the conjecture is over 70 years old, it is still undecided. It has been proved, with the aid of computers, only for games with up to 30 discs.

So what about psychology?

Psychologists have been using the Tower of Hanoi for some time, in particular for assessing patients' ability to plan ahead and chop a task into smaller chunks. The game is easy to explain and you can watch a person think, following their every step. It is also easy to vary the game. You can add more discs or stipulate new starting and goal configurations that don't have all the discs stacked up on one peg – this makes the task considerably harder.

But exploiting the game's full potential requires some mathematical expertise. This is where the collaboration of mathematicians with psychologists stems from. It has resulted in a clinical tool for assessing victims of, for example, dementia or stroke, to see which areas of the brain have been impaired.

Hanoi graphs have proved particularly popular with psychologists: a person's moves can be traced on them, seeing how their effort compares to the optimal strategy and to other patients' moves, and marking steps that caused them problems – a task that is virtually impossible without the graphs. What is more, the graphs can provide clear answers to otherwise tricky questions. For example, since every Hanoi graph is connected – there is a path between any two of the dots – you know that the puzzle can always be solved, no matter which starting and goal configurations of discs you use.

We leave you with a problem to puzzle over. Show that the smallest number of moves needed to solve the puzzle on three pegs with n discs is $2^n - 1$. Note that this number increases very quickly as n grows. Had you started 50 years ago to play the game with, say, 64 discs, moving them at a rate of one disc per second, day and night without a break, you still wouldn't have got around to moving disc 32. And don't believe that 32 is halfway to 64!

• •

FURTHER READING

[1] A.M. Hinz, S. Klavžar, U. Milutinović, and C. Petr (2013). *The Tower of Hanoi: Myths and maths.* Springer, Basel.

Source
An earlier version of this article appeared in *Plus Magazine* (<http://plus.maths.org>) on 16 November 2012.

CHAPTER 23

Career: A sample path

PHILIP HOLMES

career (noun): an occupation undertaken for a significant period of a person's life and with opportunities for progress ... (verb): move swiftly and in an uncontrolled way in a specified direction.

Apple Dictionary, 2005–2011

Biographical note

Philip Holmes was born in Lincolnshire, UK, in 1945 and attended the Universities of Oxford and Southampton. He taught at Cornell from 1977 to 1994, and since then has been at Princeton, where he is currently Eugene Higgins Professor of Mechanical and Aerospace Engineering, Professor of Applied and Computational Mathematics, Associated Faculty in the Mathematics Department, and a member of Princeton's Neuroscience Institute. Much of his research has been in dynamical systems and their applications in engineering and the physical sciences, including solid and fluid instabilities, turbulence, and nonlinear optics. In the past 15 years he has increasingly turned to biology. He currently works on the neuromechanics of animal locomotion and the neurodynamics of decision making. He has also published four collections of poems. Almost all of this came about by chance. Here are five examples.

Why I stayed in science

Like many sensitive teenagers who think themselves misunderstood, I began to write terrible verse. Arriving at Oxford in 1964 to study Engineering Science, I immediately joined the Poetry Society, members of which roundly criticised my work but failed to discourage me. I would have probably transferred to English, had the British educational system allowed it. I took long walks around the narrow, foggy streets and across Port Meadow, continued writing, remained in engineering, and took an undistinguished degree. I received many rejection slips, but finally some poems were accepted for *New Measure*, a small magazine that Peter Jay had founded with the hope of identifying new writers whose books he might subsequently publish. This led to my first collection, *3 Sections of Poems*: a literal title, perhaps appropriate for an engineer. Peter still runs Anvil Press (Greenwich, UK), and has published three more of my books.

Irrational Fraction

When it seems that part of my life
    must have belonged to another, someone
        more suitable, deserving what notice

has fallen my way and knowing precisely what to do
    next; it's as if I woke half way here, to days
        couched in these alien symbols

with which I might pretend a certain facility,
    adept at seeming to manage them,
        but not (it now becomes clear) well;

for I find myself walking on stage to applause
    from the darkened hall, bowing, and about to sit
        at the open instrument, innocent of every note.

From *Lighting the Steps: Poems 1985–2001*,
Anvil Press, London, 2002.

How I started in dynamical systems

In 1973 while writing a PhD thesis in the Institute of Sound and Vibration Research (ISVR), I noticed a small poster advertising a course on catastrophe theory (aka singularities of differentiable mappings, see also Chapter 4), to be taught by David Chillingworth. How could I resist it? I found the Mathematics Tower, sat down near the back of the room, and was soon asking a classmate, David Rand, what on earth were diffeomorphisms, k-jets, and the implicit function theorem. At that time many of Southampton's mathematicians were called David: a lack of uniqueness which may cause confusion later. I needed to learn some mathematics, and David wanted to branch into applications. We thought that a good route to these goals should begin with a joint paper. We found a preprint by Christopher Zeeman on Düffing's equation that contained a small but interesting mistake, corrected it, and sent our draft to Christopher, who replied with encouragement and excellent suggestions. He subsequently hired David (R), who is still at Warwick, while David (C) remained at Southampton. Working with the Davids on this and other problems led to my abiding interest in nonlinear dynamics.

Why I emigrated to the US

With degrees in engineering and some papers on dynamical systems in preparation, I applied for lectureships in the UK. Engineers found my interests too abstract; mathematicians asked, 'How could I teach calculus without a PhD in mathematics?' Fortunately I had met several American mathematicians and engineers during postdoctoral studies at the ISVR. They arranged a cross-continental job hunt in the fall of 1976 that led to job offers from MIT and Cornell. This trip would have been much more difficult, with our two small children, had my wife not been American. Ruth and I met at Kibbutz Zikim late in 1969. Why was I in Israel then? I had started walking from Salzburg to India in June and taken a detour, following a suggestion picked up from a fellow traveller at a cheap hotel in Istanbul. By November, Anatolia was cold and Zikim seemed a good idea. Little did I know, at that time, how good it would prove to be.

How I got into neuroscience and locomotion

One day in 1980, as an Assistant Professor in Theoretical and Applied Mechanics at Cornell, I was waiting to use the copy machine in the maths department. The person in front of me had a stack of curious graphs, about which I simply had to ask. She told me that they were extracellular recordings of action potentials of motor neurons from ventral roots of an isolated, deafferented preparation of lamprey spinal cord, and that she was Avis Cohen. She delivered a short lecture that explained some of these words, after which I asked her to make two sets of copies. It had struck me that coupled oscillators might provide a model for the lamprey's central pattern generator. My colleague Richard Rand injected some biomathematical wisdom, and so began a journey into the neuromechanics of swimming and running.

How I got deeper into neuroscience and cognitive psychology

By 1998 I'd been working on low-dimensional models of turbulence for 15 years or so (this had started at Cornell, with John Lumley and our students Nadine Aubry, Emily Stone, and Gahl Berkooz). Everybody knows that turbulence is very hard – the famous physicist Werner Heisenberg dropped it after his PhD and invented quantum mechanics. Out with the dog one morning I fell in with a neighbour and his dog (in truth, the dogs belonged to our wives). He turned out to be another Cohen (Jon). We began to talk; he told me he was building neural network models of cognition; I said 'Aha: dynamical systems and bifurcations!' This connection has kept me busy for the last 16 years.

Conclusion

Mathematics applies to many things, but it does not always help one to plan ahead. I am still having fun, and don't really know where I'm going. It is quite difficult to advise students on career choices.

Acknowledgements

Many thanks to Peter Jay and Anvil Press; to Bob White and Brian Clarkson, my PhD and postdoctoral advisors, who took a chance at the ISVR; and to Earl Dowell for calling Frank Moon at Cornell, who in turn persuaded Chairman Y. H. Pao to invite me for an interview.

· ·

FURTHER READING

[1] Philip Holmes (1986). *The green road*. Anvil Press, London.
[2] Philip Holmes (2002). *Lighting the steps: Poems 1985–2001*. Anvil Press, London.

CHAPTER 24

Sweets in the jar

STEVE HUMBLE

Imagine that you are at a summer fair which is raising funds for a big charity. As you walk through the fair you come across a stall with a large jar of sweets on it. The rules are simple. You pay a pound to enter, and for this you are allowed one chance to guess the number of sweets in the jar. At the end of the day, all of the guesses are examined, and the winner is the one with the closest guess to the number of sweets. The reward, a fine teddy bear. Now the question is, suppose that the sweets are all the same (perfect spheres). Is there any way you can estimate the correct number of sweets in the jar? At a more practical level, as a mathematician, could you advise a grain farmer and help her to work out how many grains there are in a grain silo, or a grocer on how to stack their oranges on a shelf?

The maths behind packing objects is both deep and important and leads to many difficult and challenging problems. It has a wide range of applications, from the design of integrated circuits to (remarkably) the construction of the error-correcting codes used to transmit information across the distant reaches of space. It is of extreme importance to chemists interested in how molecules are arranged in crystals to minimise their energy and is also of interest to those studying social insects living in close proximity in a hive.

The question of finding the most efficient ways to pack objects in a container was first studied by Johannes Kepler (famous for his laws on planetary motion) in 1606, after being asked by no less a person than Sir Walter Raleigh (of tobacco and cape fame) about the stacking of cannonballs on the decks of his ships. The mathematical history of packing objects has been found to be much harder than anyone had anticipated. For example, if you want to pack oranges (or indeed cannon-balls) into a box, what way of packing will get the most in? Should we pack them so that each orange is exactly on top of the one below, or should we put an orange into the gap between the oranges below, giving the familiar diamond (or face-centred cubic) packing arrangement found in a grocer's? Or maybe it is simplest and best to just throw in the oranges at random. Surprisingly, the answer to these questions is very difficult.

Kepler conjectured in 1611 that the grocers had got it right but wasn't able to prove it. The problem was then studied by many other famous mathematicians, including the great Carl Friedrich Gauss. The puzzle was solved only in 1998 by Thomas C. Hales and it turns out that the grocers (and Kepler) got it right. The face-centred cubic packing really is the best way to pack oranges so that you get the most into a box. In fact if you have a large number of spherical oranges (or sweets) in a box and you pack them in this way, then they take up 74% of the available space. The exact proportion P is

$$P = \pi / \sqrt{18}.$$

Unusually for a piece of mathematics, this result made it into the *New York Times*!

An application of this mathematics is in the subject area now called *granular matter*. It covers a vast range of processes, from the packing of domestic products to industrial processes involving the movement of grains and pellets. A greater understanding of how granular matter moves, twists, spins, and breaks is the key to how cost savings can be made during the production process. Yet granular mathematics is still not fully understood and is an area of ongoing research and development.

The link between this and error-correcting codes is less obvious but perhaps even more important. Error-correcting codes are vital for transmitting information over the Internet and the many other communication channels used every second of our lives. Suppose that a satellite takes a photo of a distant planet and wants to transmit this back to Earth in such a way that the picture does not get distorted, despite the noise and interference on the way. An error-correcting code transforms your original message (e.g. the image, represented by numbers representing the colour of each pixel) into one that is less vulnerable to corruption: when you decode the message on the other side you have a good chance of recovering something very close to the original image.

This is done using a range of mathematical ideas. One important aspect is to construct a code in which the various elements are as different from each other as possible so that an element that is corrupted still remains distinguishable from the other elements of the code.

So how does this relate to sweets? You can think of an element of the original code as the centre of the sweet. The rest of the sweet represents how much this element can be changed whilst still staying part of the same sweet, that is, while still being recognisable as the same element. Think of the size of the jar as the amount of computer memory you have available for your message. The higher the packing density of our sweets (the better the arrangement of your message elements), the more sweets we can fit into a box of the same size, that is, a more information we can convey in a message of a given size.

Let's return to the original question of counting the number of sweets in the jar. To do this we will assume that the sweets are small and spherical, and they have been shaken up in the jar so that they are packed into it as efficiently as possible. If each sweet has radius r then the volume V of each sweet is given by

$$V = \frac{4}{3}\pi r^3.$$

Now, suppose that we can estimate the volume of the jar itself. Most jars are cylindrical. If the jar has radius R and height H then its volume J is given by

$$J = \pi R^2 H.$$

The number of sweets in the jar can then be calculated by using Kepler's formula. The amount of volume that can be packed with the sweets is $0.74J$, so the number N of sweets in the jar is given by the estimate

$$N = 0.74\frac{J}{V}.$$

Hooray! Now claim your prize. Maybe we should call the teddy Johannes.

Source

An earlier version of this article appeared as a Dr Maths *Evening Chronicle* newspaper column on 10 February 2012.

CHAPTER 25

Mary Cartwright

LISA JARDINE

In his *A mathematician's apology*, published in 1940, the great mathematician G. H. Hardy argued emphatically that pure mathematics is never useful. Yet at the very moment he was insisting that – specifically – 'real mathematics has no effect on war', a mathematical break-through was being made which contributed to the wartime defence of Britain against enemy air attack. What is more, that breakthrough laid the groundwork – unrecognised at the time – for an entire new field of science.

In January 1938, with the threat of war hanging over Europe, the British Government's Department of Scientific and Industrial Research sent a memorandum to the London Mathematical Society appealing to pure mathematicians to help them solve a problem involving a tricky type of equation. Although this was not stated in the memo, it related to top-secret developments in radio detection and ranging – what was soon to become known as 'radar'.

Engineers working on the project were having difficulty with the erratic behaviour of very high-frequency radio waves. The need had arisen, the memo said, for 'a more complete under-standing of the actual behaviour of certain assemblages of electrical apparatus'. Could any of the Mathematical Society's members help?

The request caught the attention of Dr Mary Cartwright (Fig. 25.1), a lecturer in mathematics at Girton College, Cambridge. She was already working on similar 'very objectionable-looking differential equations' (as she later described them). She brought the request to the attention of her long-term colleague at Trinity College, Professor J. E. Littlewood, and suggested that they combine forces. He, she explained in a memoir written later in her life, already had the necessary experience in dynamics, having worked on the trajectories of anti-aircraft guns during the First World War.

Fig 25.1 Dame Mary Lucy Cartwright (1900–1998).
© *South Wales Evening Post.*

1

Photograph taken by John Adam, Old Dominion University

2

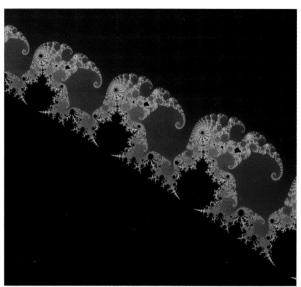

Image produced by Philip Dawid, using the program winCIG Chaos Image Generator developed by Thomas Hvel, © Darwin College, University of Cambridge

3

4

Image produced by Philip Dawid, using the program winCIG Chaos Image Generator developed by Thomas Hvel, © Darwin College, University of Cambridge

5

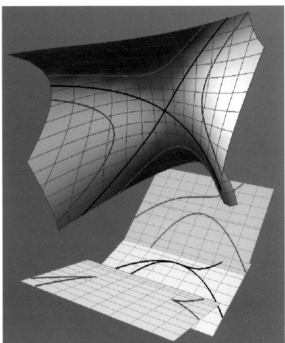

Photograph taken by John Adam, Old Dominion University

7

Created by Peter Giblin and Gordon Fletcher. © Peter Giblin

6

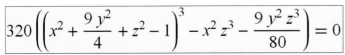

$$320\left(\left(x^2 + \frac{9\,y^2}{4} + z^2 - 1\right)^3 - x^2\,z^3 - \frac{9\,y^2\,z^3}{80}\right) = 0$$

Created by Michael Croucher using Mathematica™, first published at <www.demonstrations.wolfram.com/EquationsForValentines/>, © Wolfram Demonstrations Project & Contributors 2013

8

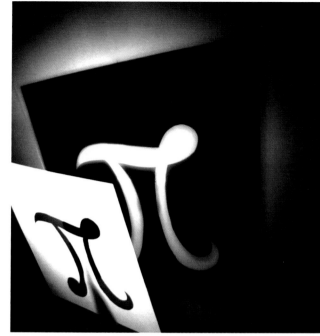

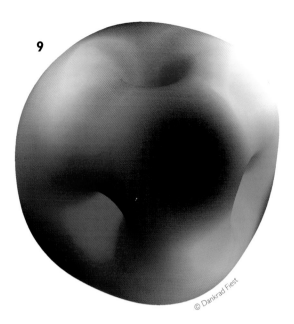

9

© Dankrad Fiest

10

11

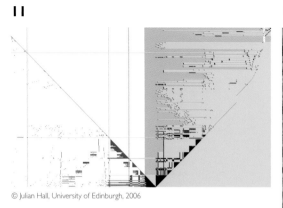

12

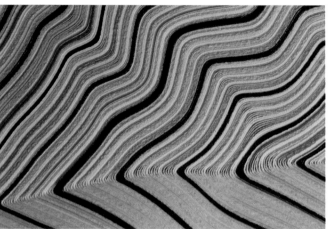

Created by Timothy Dodwell and Andrew Rhead of the Composite Research Unit, University of Bath

13

14

15

16

17

Created by R. R. Hogan, University of Cambridge

18

19

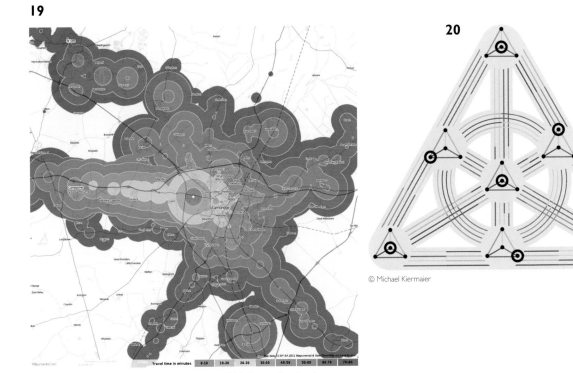

Created by Mapumental using Open Street Map and the National Public Transport Data Repository.
Creative Commons license CC-BY-SA 2011

20

21

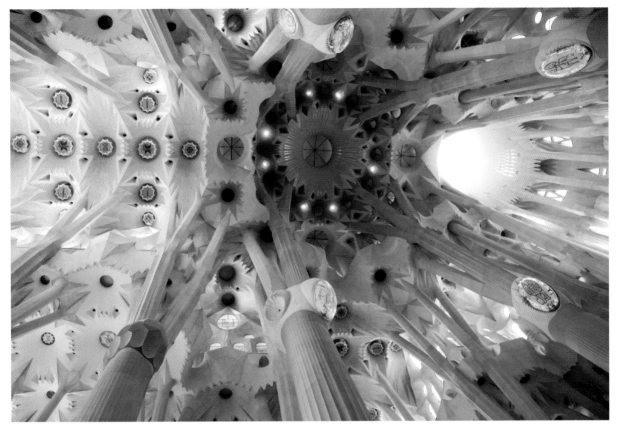

22

23

Image created by Benjamin Favier, University of Cambridge, using VAPOR (<www.vapor.ucar.edu>) from The US National Center for Atmospheric Research

24

Created by Oliver Labs, using the software Singular and visualised using Surf. © Oliver Labs, <www.MO-Labs.com>

25

26

27

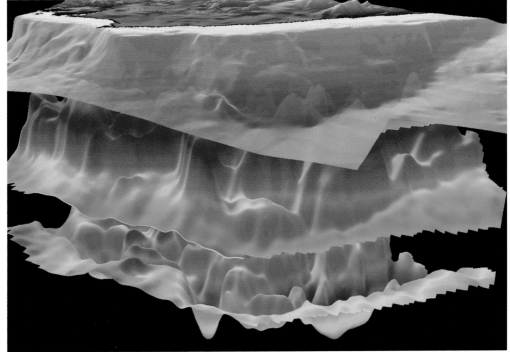

28

29

30

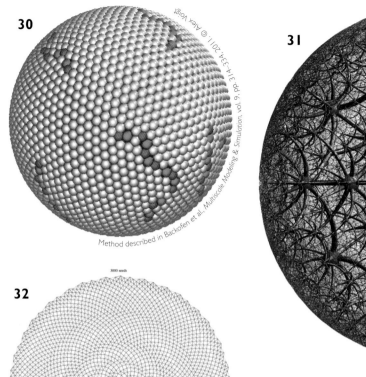

Method described in Backofen et al., *Multiscale Modeling & Simulation*, vol. 9, pp. 314–334, 2011. © Alex Voigt

31

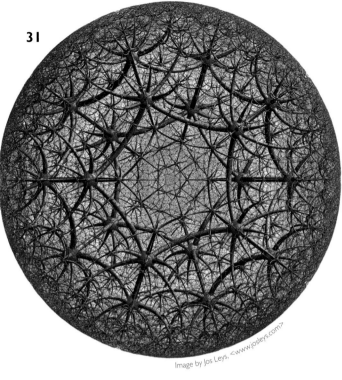

Image by Jos Leys, <www.josleys.com>

32

3000 seeds

© Ron Knott

33

© Ben Laurie

34

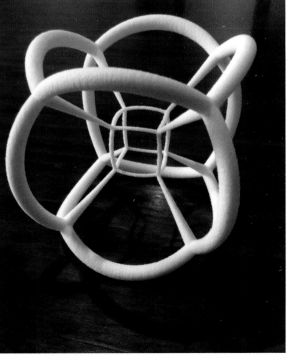

Image by Henry Segerman of a sculpture made by Saul Schleimer and Henry Segerman

35

Created by Martin Schneider using the software available at
<www.openprocessing.org/sketch/7579>

36

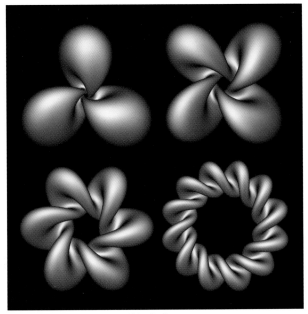

Created by Nicholas Schmidt, Tübingen University, Germany

37

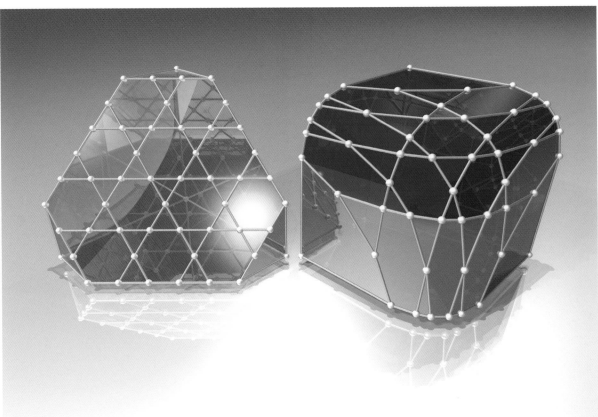

38

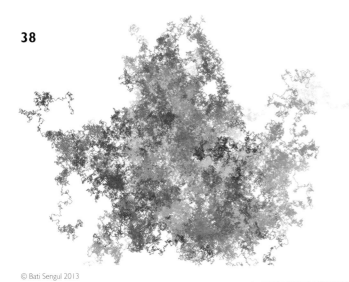

39

40

41

42

43

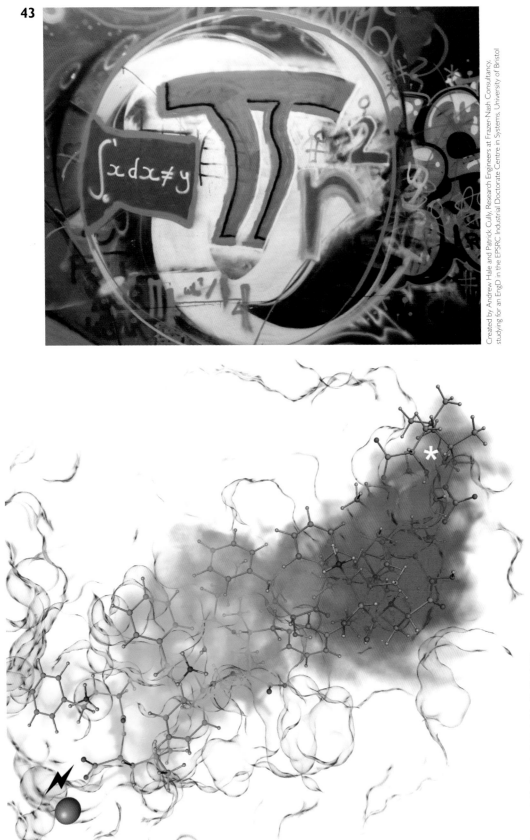

Created by Andrew Hale and Patrick Cully, Research Engineers at Frazer-Nash Consultancy, studying for an EngD in the EPSRC Industrial Doctorate Centre in Systems, University of Bristol

44

Image created by Alexander Bujotzek, Peter Deuflhard, and Marcus Weber. © Zuse Institute Berlin, Germany

45

Photograph courtesy of the Department of Civil and Environmental Engineering, Imperial College London

46

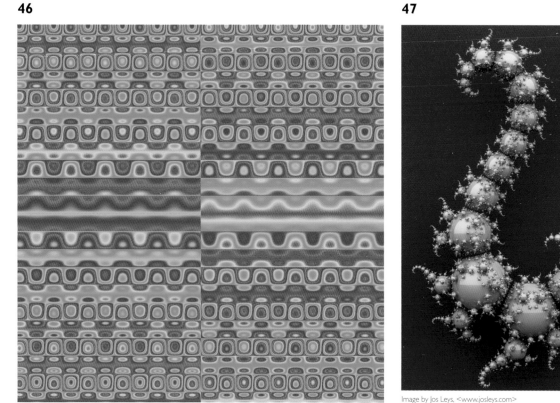

Created by Maksim Zhuk using the Pygame library within Python

47

Image by Jos Leys, <www.josleys.com>

48

Image by Julian Landel, Colm Caulfield, and Andy Woods from *Journal of Fluid Mechanics*, vol. 692, pp. 347–368
© Cambridge University Press

49

50

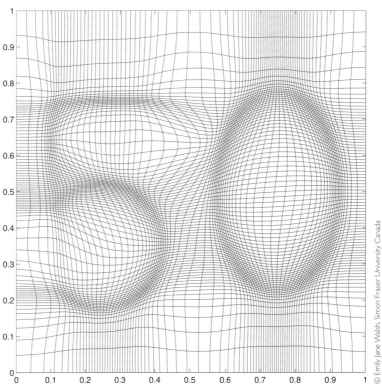

The distinguished physicist and public intellectual Freeman Dyson (see Chapter 17) – who is British by birth, but has, since the 1950s, spent most of his professional life at the Princeton Institute for Advanced Studies in America – heard Cartwright lecture on this work when he was a student at Cambridge in 1942. He has given us a vivid account of the importance of the war work Cartwright and Littlewood did:

The whole development of radar in World War Two depended on high power amplifiers, and it was a matter of life and death to have amplifiers that did what they were supposed to do. The soldiers were plagued with amplifiers that misbehaved, and blamed the manufacturers for their erratic behaviour. Cartwright and Littlewood discovered that the manufacturers were not to blame. The equation itself was to blame.

In other words, odd things happened when some sorts of values were fed into the standard equation they were using to predict the amplifiers' performance. Cartwright and Littlewood were able to show that as the wavelength of radio waves shortens, their performance ceases to be regular and periodic, and becomes unstable and unpredictable. This work helped explain some perplexing phenomena engineers were encountering.

Cartwright herself was always somewhat diffident when asked to assess the lasting importance of her war work. She and Littlewood had provided a scientific explanation for some peculiar features of the behaviour of radio waves, but they did not in the end supply the answer in time. They simply succeeded in directing the engineers' attention away from faulty equipment towards practical ways of compensating for the electrical 'noise' – or erratic fluctuations – being produced.

So while Cartwright and Littlewood were producing significant results on the stability of solutions to the equation describing the oscillation of radio waves, the engineers working on radar systems decided they could not wait for precise mathematical results. Instead, once it had been identified, they worked round the problem, by keeping the equipment within predictable ranges.

Perhaps in part because of her own overly modest assessment of its importance, Cartwright's original work went relatively unnoticed when it was published in the *Journal of the London Mathematical Society* shortly after the end of the war. Freeman Dyson maintains that this is a classic example of the way in which real mathematical originality and innovation is missed until a generation after the work has been done:

When I heard Cartwright lecture in 1942 I remember being delighted with the beauty of her results. I could see the beauty of her work but I could not see its importance. I said to myself, 'This is a lovely piece of work. Too bad it is only a practical wartime problem and not real mathematics.' I did not say, 'This is the birth of a new field of mathematics.' I shared the tastes and prejudices of my contemporaries.

The 'new field' Dyson refers to here, which he and his contemporaries failed to recognise, is chaos theory. Cartwright's early contribution to the field is now acknowledged in all histories of the subject, but was largely overlooked for almost 20 years. The results unexpectedly obtained from the equations predicting the oscillations of radio waves are part of the foundation for the modern theory which accounts for the unpredictable behaviour of all manner of physical phenomena, from swinging pendulums and fluid flow to the stock market. Steadily increase the rate of flow of water into a rotating water-wheel, for example, and the wheel will go correspondingly faster. But at a certain point the behaviour of the wheel becomes unpredictable – speeding up and slowing down without warning, or even changing direction.

The recognition that chaotic behaviour is a vital part of many physical systems in the world around us came in 1961, when Lorenz was running a weather simulation through an early computer. When he tested a particular simulated configuration a second time he found that the outcome differed dramatically from his earlier run. Eventually he tracked the difference down to a small alteration he had inadvertently made in transferring the initial data, by altering the number of decimal places. Lorenz immortalised this discovery in a lecture entitled *Does the flap of a butterfly's wings in Brazil set off a tornado in Texas?* Today, when we think of chaos theory we associate it with all kinds of fundamentally unstable situations – but one of the most vivid to imagine is still the idea that one flap of a butterfly's wing deep in the Amazon rainforest is the cause of a weather system thousands of miles away.

This is the same kind of unpredictability arising from small changes in initial conditions that Cartwright and Littlewood had recognised and drawn attention to in their work with radio waves several decades earlier.

After the war Mary Cartwright moved away from knotty differential equations and ended her collaboration with Littlewood. She went on to have a distinguished academic career in pure mathematics and academic administration, earning a succession of honours.

In 1947 she was the first woman mathematician to be elected to the Royal Society. In 1948 she became Mistress of Girton College, Cambridge, and then Reader in the Theory of Functions in the Cambridge Mathematics Department in 1959. From 1961 to 1963 she was President of the London Mathematical Society, and received its highest honour, the De Morgan Medal, in 1968. She was made a Dame Commander of the British Empire in 1969.

She lived long enough to see the field in which she had made those early, important discoveries become a major part of modern mathematics, and to see it take its place in the popular imagination. She was, however, characteristically modest to the end about the part she had played. Freeman Dyson claims that 'Littlewood did not understand the importance of the work that he and Cartwright had done. Only Cartwright understood the importance of her work as the foundation of chaos theory, and she is not a person who likes to blow her own trumpet.' He records, however, that shortly before her death, he received an indignant letter from Cartwright, scolding him for crediting her with more than she deserved.

Dame Mary Cartwright died in 1998 at the age of 97. In one of the many obituaries paying tribute to her, a friend and colleague describes her as 'a person who combined distinction of achievement with a notable lack of self-importance.' She left strict instructions that there were to be no eulogies at her memorial service.

However, now 70 years after her first seminal achievement, it seems an appropriate time to blow Dame Mary Cartwright's trumpet on her behalf – for her brilliance as a mathematician, and as one of the founders of the important field of chaos theory.

. . . .

FURTHER READING

[1] G. H. Hardy (1940). *A mathematician's apology*. Cambridge University Press.
[2] Freeman Dyson (2006). Mary Lucy Cartwright [1900–1998]: Chaos theory. In: *Out of the shadows: Contributions of twentieth-century women to physics*, edited by Nina Byers and Gary Williams, pp. 169–177. Cambridge University Press.

Source

The content of this article first appeared in a BBC Radio 4 programme broadcast on 8 March 2013.

CHAPTER 26

The fallibility of mathematics

ADAM JASKO

When I was young, and indeed to this day, I loved experimenting to discover interesting results or perhaps to verify those already known to me. In the sciences this involves getting your hands dirty: mixing various chemicals to observe a reaction, growing runner beans in jam jars and comparing those placed in cupboards to those under sunlight, rolling trolleys down ramps to find the acceleration due to gravity, and so on. This inevitably leads on to experiments that can't be done at home, for example imaging complex body tissue or trying to find the Higgs boson. These experiments require very expensive and often very large equipment such as MRI scanners or the Large Hadron Collider at CERN. In maths, however, the only dirtying of one's hands required is from the ink of a biro.

Often in the sciences one must simply believe a teacher or lecturer when they say that atoms are made of protons, neutrons, and electrons, or that cells contain a double-helix structure that encodes all your genetic information. In maths, however, you need not take their word for it. If you do not believe some trigonometric identity or the answer to a large sum then you can just check it for yourself. One of the attractive things about maths is that it can be done almost anywhere and the only accessories required are pen and paper.

Another thing that separates maths from the other sciences is that once some result has been proved, there is no need to recheck. Scientific theories are constantly under scrutiny and may at any point become inadequate at explaining some newly observed events and need to be replaced with more fitting alternatives. Mathematical truth, however, doesn't change with time.

Of course, to have this mathematical guarantee we need to be sure our initial reasoning is correct. Take the following proof. Let a and b denote the same arbitrary number. Then

$a = b$	basic assumption
$a^2 = ab$	multiply both sides by a
$2a^2 = a^2 + ab$	add a^2 to both sides
$2a^2 - 2ab = a^2 - ab$	subtract $2ab$ from both sides
$2(a^2 - ab) = 1(a^2 - ab)$	factorise the left-hand side
$2 = 1$	cancel the common factor.

Some may look at this and jump to the conclusion that 2 really is equal to 1 (indeed, I once used this argument to convince a friend to doubt all maths), yet it soon becomes apparent that this is no proof at all. Upon inspection it should become clear that the last step contains a fallacy, namely dividing by zero. Even this simple trick has the ability to deceive.

Other proofs are far more subtle and mistakes are less noticeable than in the above, so how can we be sure that an error has not been made? The short answer is that we can't, and many mathematicians have published proofs in which a flaw was found, sometimes at a far later date. The scrutiny of the academic community can be as important to help confirm mathematical theorems as it is to verify scientific theories.

One big problem is big proofs. Some theorems are so large that they become too hard and laborious to follow for most. The perfect example is the classification theorem, which I shall not explain here, but which involved finding all *finite simple groups* (these objects can be thought of as the basic building blocks in an area of mathematics called *group theory*) and showing that no others exist. This massive proof spans about 500 articles over half a century. The mathematics is complicated enough to follow, and when one considers that very few have gone through the tens of thousands of pages one can start to question one's faith in the result.

With good reason too: in 1983 the proof was announced to be complete, but a bit later a mistake was found. This was eventually corrected but it was not until 2008 that a revised version was published. Even now doubts can carry on lurking, and Michael Aschbacher (one of those responsible for making those final adjustments to the theorem) said that 'The probability of an error in the proof of the classification theorem is virtually 1. On the other hand the probability that any single error cannot be easily corrected is virtually zero, and as the proof is finite, the probability that the theorem is incorrect is close to zero.'

Of course, one could argue from a pedantically philosophical perspective that no matter how many mathematicians go through the proof with a fine-tooth comb we can never be sure that there is not some small error that everyone has missed. Perhaps Pythagoras's theorem has some flaw that nobody has yet noticed. This kind of thinking leads us nowhere, though, if we desire to seek any truths at all. Such fancies aside, there are still important issues with the truth and wholeness of mathematics.

Kurt Gödel's (first) *incompleteness theorem*, published in 1931, provides an insight into proof and mathematical systems. This result states that all sufficiently powerful *axiomatic systems* are either *incomplete* or *inconsistent*. An axiomatic system is based on a set of axioms (self-evident truths) and rules to make logical inferences. Gödel's incompleteness theorem states that if such a system is powerful enough to express the arithmetic of the whole numbers and is free from contradictions, then it can express statements that cannot be proven true or false within the system (see Chapter 31 for more details).

This is rather shocking, and you may wonder why Gödel's result has not wiped out mathematics once and for all. The answer is that the unprovable statements logicians have found so far do not touch on ordinary everyday mathematics. They certainly would not come up in school homework or in the maths used to design aircraft wings. The vast majority of mathematicians leave these logical holes to the logicians and philosophers and get on with their work.

With the dawn of technology, mathematics has evolved down routes that could not have been foreseen. Enormous numerical computations can be performed and computers have begun to play a role in proofs, most notably that of the famous *four colour theorem*. This old problem asks whether it is always possible to colour a map, using no more than four colours, so that no two bordering countries are coloured the same. After many fallacious proofs, the four colour theorem was eventually proved (four colours do indeed suffice), but the proof relied heavily on computers to check through a large number of possible configurations. This approach has been questioned, since it is impossible to check the computer's calculations by hand. Consequently, some mathematicians remain sceptical of proofs heavily reliant on computers.

In conclusion, complicated (and perhaps unpublished) programs run on supercomputers do not offer the same accessibility as pen and paper proofs, especially to the interested amateur. The length and technicality of other theorems can make verification difficult and even mathematicians can make mistakes. So although in mathematics infallible truth is in theory achievable, the skills, time, understanding, and in some cases computing power required to verify certain results for oneself can render it as inaccessible as a Large Hadron Collider. There is no need to worry about Pythagoras's theorem, though: it can be proved quite easily on pen and paper so Pythagoras, at least, is safe!

. . . .

FURTHER READING

[1] Keith Devlin (1998). *Mathematics: A new golden age*. Penguin (2nd edition).
[2] Douglas Hofstadter (1980). *Gödel, Escher, Bach: An eternal golden braid*. Penguin.

Source
Selected as an outstanding entry for the competition run by *Plus Magazine* (<http://plus.maths.org>).

CHAPTER 27

Anecdotes of Dr Barrow

TOM KÖRNER

There have been few longer-lasting and ultimately futile controversies than that over the invention of calculus. At the beginning of the 20th century, J.M. Child sought to bring discussion to a close with the uncompromising statement that 'Isaac Barrow was the first inventor of the Infinitesimal Calculus'. Who was Isaac Barrow and why should such a claim be remotely plausible?

Historians talk about the unremembered lives of common people, but we know very little about the lives of many uncommon people. Isaac Barrow was fortunate to find a place in Aubrey's *Brief Lives*. That a mathematician should be described as 'by no means a spruce man, but most negligent in his dresse' comes as no surprise, but we also learn that 'he feared not any man' and as a student 'He would fight with the butchers' boyes . . . , and be hard enough for any of them.'

Barrow belonged to a family ruined in the support of the King against Parliament in the English Civil War. His obvious ability and personal charm secured him sufficient support for an excellent education, whilst his outspoken royalism remained a constant worry for his friends.

He became a fellow of Trinity College, Cambridge, but the college, anxious to keep him out of trouble, paid for him to travel through Europe for several years. He studied modern mathematics in Paris (by no means the last Cambridge mathematician to do so), met Galileo's last pupil in Florence, fought pirates who attacked his ship in the Mediterranean, visited Constantinople, and lost his entire baggage in a ship fire in Venice.

When he returned, the King had been restored and the University appointed him Professor of Classics. He filled the post with distinction (he spoke six languages and wrote poetry in English and Latin), producing new editions of Euclid, Archimedes, and Appollonius. However, as his choice of ancient texts shows, he was more interested in mathematics and he moved to the better-paid post of Lucasian Professor when it was established.

Barrow's splendidly titled *The usefulness of mathematical learning explained and demonstrated* begins with a chapter of praise for Lucas, 'A *Mecænas* not in Name, but in Fact; not one who has made a shew of bare favour towards learning, but who has spent much real labour on it; not who has embraced it with a good Will only, but who has obliged it with a Munificent hand . . . inspiring Vigour into these languishing studies.' Those who find the oratory overdone should reflect that, apart from Barrow himself, occupants of the Lucasian chair have included Newton, Airy, Babbage, Stokes, Dirac, and Hawking. It seems likely that Barrow's lectures were the only Cambridge lectures from which Newton, who was otherwise self-taught, would have benefited.

Soon Barrow was writing about 'Mr. Newton a fellow of our College and very young . . . but of an extraordinary proficiency in these things.' He gave Newton the run of his library and tried to persuade him to publish his discoveries. When Barrow decided that theology was still more important than mathematics and resigned his chair, he probably recommended Newton to be his successor.

Barrow soon proved himself one of the greatest preachers in England and was appointed Chaplain to the King. His natural wit and pugnacity are shown in a well-known exchange with Lord Rochester, a man as famous for his wickedness as Barrow was famous for his goodness:

Rochester, bowing deeply: Doctor, I am yours to the shoetie.
Barrow, bowing still more deeply: My lord, I am yours to the ground.
Rochester: Doctor, I am yours to the centre.
Barrow: My lord, I am yours to the antipodes.
Rochester: Doctor, I am yours to the lowest pit of hell.
Barrow, turning on his heel: There, my lord, I leave you.

The King may not have viewed the presence of such a clever, religious, and outspoken man as an unalloyed blessing. In any case, he appointed Barrow to the Mastership of Trinity, saying that he gave it to 'the best scholar in England.' Here Barrow performed his last and perhaps most important service to Newton. Fellowships at Trinity demanded ordination, something which would have been incompatible with Newton's private, but strongly held, religious views. Barrow secured exemption for the holders of the Lucasian chair and so freed Newton to pursue his research.

As Master of Trinity, Barrow served on a committee charged with finding a central meeting place for the University. There he argued that 'If they made a sorry building, they might fail in contributions; but if they made it very magnificent and stately, and at least exceeding Oxford, all gentlemen of their interest would generously contribute.' When he failed to convince the committee, he determined to show what a single college could do, and persuaded Trinity to build and his friend Wren to design a splendid library.

The cost of the new building put a severe strain on the College's finances, but the library remains one of the most splendid buildings in Cambridge. Mathematical tourists will note, among the statues adorning the outside, the figure of Mathematics herself, apparently counting on her fingers.

For 300 years after his death Barrow's reputation rested on his sermons and works of religious controversy, although his mathematical ability was not forgotten.

Child's claims for Barrow's invention of calculus are based on Barrow's *Geometrical Lectures*. In it Child finds 'a complete set of standard forms for both the differential and integral sections of the subject', the standard rules for differentiation (including the quotient rule, but not the chain rule), the calculus treatment of tangents ('Barrow's differential triangle'), and a clear statement and proof of the fundamental theorem of calculus.

What do modern historians make of this? It must be said that there seem to be as many opinions as there are historians. I am reminded of Borge's account (see, e.g. <http://www.pen.org/nonfiction-transcript/all-range>) of how

working with enthusiasm ... through the English version of a certain Chinese philosopher, I came across this memorable passage: 'A man condemned to death does not care that he is standing at the edge of a precipice, for he has already renounced life.' Here the translator attached an asterisk, and his note informed me that this interpretation was preferable to that of a rival Sinologist, who had translated the passage thus: 'The servants destroy the works of art, so that they will not have to judge their beauties and defects.'

What I see as consensus may simply be what I wish to see.

The first point to make is that what Child saw as there is actually there, though in a geometric form which seems strange to modern eyes. It is no accident that the influential Russian mathematician Vladimir Arnol'd (1937–2010), who wished to 'regeometrise' calculus, was also a great admirer of Barrow. The second is that, just as Child saw much of Newton and Leibniz in Barrow, so modern historians see much of Barrow in his predecessors. Earlier mathematicians had some inkling of the fundamental theorem even if Barrow was the first to write it down in full generality.

We can write down a long list of mathematicians, including Archimedes, Galileo, Kepler, Cavalieri, Sluse, Fermat, Pascal, Descartes, Wallis, and Gregory, whose work was known directly and indirectly to Newton and Leibniz and who may be viewed as forerunners of calculus. The questions Newton and Leibniz asked and the solutions they gave could not have been produced one hundred or even fifty years before their time. (On the other hand, I do not think it is possible to name a *single* mathematical result which Newton could not have obtained for himself.)

However, there are good reasons why we talk of 'forerunners' and 'precursors' and not inventors. Before Newton and Leibniz, calculus was a collection of problems and methods, usually cast in geometric form. After Newton and Leibniz, it was a single coherent subject, cast in algebraic and algorithmic form, which provided an unparalleled tool for the description of the physical Universe. Leibniz was only exaggerating slightly (and, so far as mathematical physics goes, not exaggerating at all) when he said 'Taking mathematics from the beginning of the world to the time of Newton, what he has done is much the better half.' The successors of Newton and Leibniz had no reason to look at earlier work and did not do so.

Would Barrow have minded? I doubt it. He does not appear to have cared greatly about the publication of his mathematical work and he would certainly have considered his theological work – in particular his defence of the doctrines of the Church of England – as the true centre of his life. Aubrey relates that, on his deathbed, 'the standers-by could hear him say softly "I have seen the glories of the world." ' The man who built Trinity Library and taught Newton had little to regret.

CHAPTER 28

Finding Apollo

ADAM KUCHARSKI

When NASA first decided to put a man on the Moon they had a problem. Actually, they had several problems. It was the spring of 1960 and not only had they never sent a man into space, the Soviet Union had recently won the race to put a satellite in orbit. Then there were the technical issues: the nuts and bolts of how to transport three men 238,900 miles from a launch site in Florida to the Moon – and back again – all before the Soviets beat them to it.

One of the biggest stumbling blocks was estimating the spacecraft's trajectory: how could NASA send astronauts to the Moon if they didn't know where they were? Researchers at NASA's Dynamics Analysis Branch in California had already been working on the problem for several months, with limited success. Fortunately for the Apollo programme that was about to change.

Working out the trajectories had proven difficult for the researchers because they were really trying to solve two problems at once. The first was that spacecraft don't accelerate and move smoothly in real life, as they might in a physics textbook. They are subject to variable effects like lunar gravity, which are often unknown. This randomness meant that even if the scientists could have observed the craft's exact position, the trajectory wouldn't follow a neat, predictable path.

But that was their second problem: they couldn't observe its exact position. Although on-board sensors included a sextant, which calculated the angle of the Earth and Moon relative to the space-craft, and gyroscopes, it wasn't clear how these data – which inevitably contained measurement errors – could be accurately translated back into position and velocity.

Stanley Schmidt, the engineer who led the Dynamics Analysis Branch, had initially hoped to use ideas developed for long-range missiles, another product of the Cold War rivalry with the Soviet Union. However, missile navigation systems took measurements almost continuously, whereas during a busy mission the Apollo craft would only be able to take them at irregular intervals. Schmidt and his colleagues soon realised that existing methods wouldn't be able to give them accurate enough estimates: they needed to find a new approach.

What happened next was an incredible stroke of good luck. In the autumn of 1960 an old acquaintance of Schmidt's – who had no idea about the work the NASA scientists were doing – called to arrange a visit. Rudolph Kalman was a mathematician based in Baltimore and he wanted to come and discuss his latest research.

Kalman specialised in electrical engineering and had recently found a way of converting a series of unreliable measurements into an estimate for what was really going on. His mathematical results had been met with scepticism, though, and Kalman had yet to find a way to turn his theory into a practical solution.

Schmidt didn't need convincing. After hearing Kalman's presentation he distributed copies of the method to the NASA engineers and worked with Kalman to develop a way to apply it to their problem. By early 1961 they had their solution.

Kalman's technique was called a *filter* and it worked in two steps. The first used Newtonian physics to make a prediction about the current state of the system (in NASA's case the location of the spacecraft) and the level of uncertainty due to possible random effects. The second step then used a weighted average to combine the most recently observed measurement, which inevitably had some degree of error, with this prediction.

To give a simplified example, suppose a mini-spacecraft moves along the edge of a ruler between the numbers 0 and 1. Let the predicted position at a given time be 1/4 and the observed position be 3/4. If K represents the relative level of confidence in the prediction compared to the observation, the weighted average is

$$\frac{1}{4}K + \frac{3}{4}(1 - K).$$

If we were equally confident in the accuracy of the predicted and observed positions, we could assume $K = 1/2$. The weighted average would then be a simple average of the two values,

$$\frac{\frac{1}{4} + \frac{3}{4}}{2} = \frac{1}{2}.$$

The Kalman filter, however, does something cleverer than this. It takes the level of randomness that went into the prediction, and the amount of error we think there is in the measurement, and combines them to give the optimal value for the relative confidence K. If the prediction was trusted more than the measurement, more weight is given to the prediction. Conversely, if the measurements were more plausible, then these are given more preference.

An example of a weighted average which gives more weight to the predicted position than the observed one (in fact, twice as much weight) would be one where $K = 2/3$. In this case, the filter gives us an estimate closer to the predicted position,

$$\frac{2}{3} \times \frac{1}{4} + \frac{1}{3} \times \frac{3}{4} = \frac{5}{12} = 0.42,$$

as shown in Fig. 28.1.

As well as producing accurate estimates, the Kalman filter could run in real time: all it needed to generate an estimate were the previous prediction and the current on-board measurement. Because any calculations would have to be done on the Apollo's primitive on-board computer, this simplicity made the filter incredibly valuable. In fact, it would eventually be used on all six Moon landings, as well as finding its way into the navigation systems of the International Space Station.

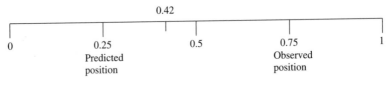

Fig 28.1 A weighted average.

Years after the Apollo programme finished, Stanley Schmidt and one of his colleagues produced a report documenting their work on the trajectory predictions. As expected, they had plenty of praise for the Kalman filter. 'The broad application of the filter to seemingly unlikely problems suggests that we have only scratched the surface when it comes to possible applications,' they wrote, 'and that we will likely be amazed at the applications to which this filter will be put in the years to come.'

It turned out to be a shrewd prediction. Disease researchers now employ filters to work out what causes outbreaks. Atmospheric scientists use them to make sense of weather patterns. Economists apply the methods to financial data.

The techniques may have begun as a way to guide three men to the Moon, but they have since evolved into a valuable set of tools for tackling problems back here on Earth.

. . . .

FURTHER READING

[1] Leonard McGee and Stanley Schmidt (1985). *Discovery of the Kalman filter as a practical tool for aerospace and industry*. NASA Technical Memorandum 86847, available from <https://ece.uwaterloo.ca/~ssundara/courses/kalman_apollo.pdf> (site accessed May 2013).

[2] Greg Welch and Gary Bishop (2006). *An introduction to the Kalman filter*. Technical Report 95-041, Department of Computer Science, University of North Carolina, available from <http://www.cs.unc.edu/~welch/kalman/kalmanIntro.html> (site accessed May 2013).

Source

A longer version of this article first appeared in *Plus Magazine* (<http://plus.maths.org>) on 11 October 2012.

CHAPTER 29

The golden ratio in astronomy and astrophysics

MARIO LIVIO

The golden ratio, commonly denoted in the popular mathematics literature by the Greek letter ϕ (phi), has a propensity for popping up where least expected. It appears in a variety of natural phenomena and works of art. In its decimal representation, ϕ is equal to the never-repeating, never-ending number which starts 1.6180339887.... Where can such an irrational number show up in astronomy and astrophysics? First, it appears in what has become one of the best-recognised astronomical symbols. The left image in Fig. 29.1 shows a pentagram, a five-pointed star, which is widely used as a symbol for a real star; the shape reflects the twinkling of stars due to the Earth's atmosphere. For each of the five congruent outer triangles of a pentagram, the ratio of the side to the base is precisely equal to ϕ. Second, when Plato wanted to discuss the cosmos as a whole in his celebrated *Timaeus*, he chose one of the five so-called Platonic solids, the dodecahedron (see the right image in Fig. 29.1), as the shape 'which the god used for embroidering the constellations on the whole heaven'. The dodecahedron has the golden ratio written all over it. If you take a dodecahedron with an edge length of 1 unit, its volume and surface area can be expressed as simple functions of ϕ.

Amusingly, Plato's choice for describing the universe as a whole has a modern reincarnation. In 2003, a group of cosmologists proposed that certain features in the observations of the 'afterglow of creation' – the cosmic microwave background – could be explained if the universe is in fact finite (positively curved), and shaped like a dodecahedron! It would have been truly amazing if

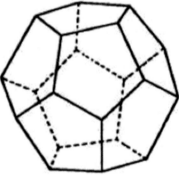

Fig 29.1 (Left) A pentagram. (Right) A dodecahedron.

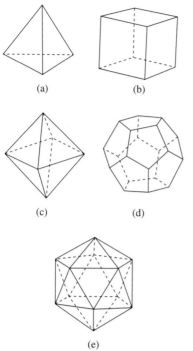

(a)   (b)

(c)   (d)

(e)   **Fig 29.2** The Platonic solids.

Plato were right after all (even though he clearly had something else in mind). However, more detailed analysis of the data from the Wilkinson Microwave Anisotropy Probe suggests that the universe is infinite and geometrically flat.

The Platonic solids shown in Fig. 29.2 play a crucial role also in a model of the solar system constructed in 1596 by the astronomer Johannes Kepler. Kepler wondered why there were six planets – only six were known at his time – and what determined the spacings between their orbits. After many different geometrical calculations, he concluded that the number of planets and their orbits were fixed by the fact that there were precisely five Platonic solids (tetrahedron, cube, octahedron, dodecahedron, and icosahedron). Embedding the Platonic solids one inside the other as shown in Fig. 29.3, together with an outer spherical boundary corresponding to the heaven of the fixed stars, defines six spacings. Kepler showed that with a particular choice for the order of the embedding, he could get the relative sizes to agree with the radii of the planetary orbits to within about 10%. The golden ratio plays a central role not just in the properties of the dodecahedron, but also in its dual solid, the icosahedron. As with the dodecahedron, both the surface area and the volume of an icosahedron with unit-length edges can be expressed as simple functions of ϕ. While Kepler's model was clearly wrong, the problem was not with the geometrical arguments. Kepler's explanation failed because he did not understand that neither the number of the planets nor their orbital radii are fundamental quantities that require explanation from first principles.

Another intriguing area of astrophysics in which the golden ratio makes an unexpected appearance is that of black holes. Black holes warp spacetime in their vicinity so much that in Einstein's classical general relativity nothing can escape from them, not even light. However,

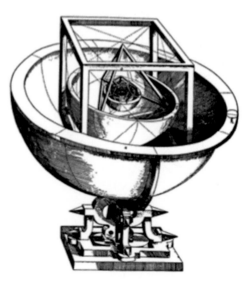

Fig 29.3 Schematic from Kepler's book *Mysterium Cosmographicum* which illustrates the model of the solar system.

when quantum effects are brought into the picture, black holes can lose energy via a process known as Bekenstein–Hawking radiation. Astrophysical black holes are characterised by two physical properties, their mass and their angular momentum, a measure of how fast they are spinning. These spinning black holes are called *Kerr black holes*, after the New Zealander physicist Roy Kerr. They can exist in two states: one in which they heat up when they lose energy, that is to say, with negative specific heat, and another in which they cool down, that is, with positive specific heat. They can also undergo a phase transition from one state into the other, in the same way that water can freeze to form ice. In 1989, the British physicist Paul Davies showed that the transition takes place when the square of the black-hole mass (in appropriate but natural units) is precisely equal to ϕ times the square of its spin (again in appropriate units). I should note that there is nothing particularly mysterious about the appearance of the golden ratio in this phase transition. It basically stems from the simple fact that the golden ratio is a solution to the quadratic equation $x^2 = x + 1$. The other is ϕ^{-1}.

Finally, it turns out that the golden ratio happens to be *optimal* for the design of certain radio telescopes. The goal of such telescopes is to place radio antennas so as to sample the visibility plane uniformly, avoiding overlapping and redundancy. The fact that the golden ratio is ideal for this task has already been 'discovered' by many plants in their leaf arrangement (the phenomenon known as phyllotaxis). The leaves along a twig of a plant tend to grow in positions that optimise their exposure to the sun, air, and rain. This is best achieved when the angle between successive leaves equals the *golden angle* of approximately 137.5°, or more precisely 360°/ϕ subtracted from a full circle. It can be shown mathematically that the golden angle allows for the most efficient filling of space without overlapping. Amazing, isn't it, for one number to connect black holes to the telescopes that observe them?

If you examine all the phenomena with explanations in which the golden ratio makes a seemingly unexpected appearance, you'll discover that they normally fall into one of three categories. First, the golden ratio is at the root of any fivefold symmetry: in any regular pentagon, the ratio of the length of a diagonal to the length of a side is equal to ϕ. The second instance has to do with the fact that ϕ is a solution to the simple quadratic equation $x^2 = x + 1$. The third category is perhaps the most interesting one. The golden ratio is the 'most irrational' of all irrational numbers, in the sense that when expressed as a continued fraction it is the slowest to converge. This stems from the fact that its continued fraction is composed exclusively of 1s:

$$\phi = 1 + \cfrac{1}{1 + \cfrac{1}{1 + \cfrac{1}{1 + \cfrac{1}{1 + \ldots}}}}.$$

The conclusion is simple. The golden ratio has many special properties, and the astonishing appearance of ϕ in so many applications can always be traced back to one of these.

CHAPTER 30

The high-power hypar

PETER LYNCH

The capacity of mathematics to provide general, unifying structures is one of its most powerful characteristics. Mathematics frequently shows us surprising and illuminating connections between physical systems that are not obviously related: the analysis of one system often turns out to be ideally suited for describing another.

To illustrate this, we will show how a surface in 3-dimensional space – the hyperbolic paraboloid, or *hypar* – pops up in surprising ways and in many different contexts. In the process we find unexpected connections between architecture, tennis balls, weather forecasting, and the snack food called Pringles.

Curves and surfaces

In two dimensions, a point (x, y) is given by the two coordinates x and y. Each is free to vary independently; we say the point has two degrees of freedom. If we now specify an equation $f(x, y) = 0$, the dimension is usually reduced by one if the equation has solutions: instead of the whole plane, we have a 1-dimensional subset, a curve. In the special case where the equation is linear, the curve is a straight line.

Moving up a notch, a point (x, y, z) in 3-dimensional space is given by three coordinates x, y, and z. If we specify an equation $g(x, y, z) = 0$, the point is confined to a 2-dimensional surface. In the special case of a linear equation, that surface is a plane in 3-dimensional space.

To describe a *curve* in 3-dimensional space, we need to reduce the dimension once more, by giving a second equation, $h(x, y, z) = 0$. If both equations are linear, they describe two planes, whose intersection is typically a straight line. More generally, they are nonlinear, and provided they intersect they describe a 1-dimensional curve.

The most ancient and best-understood curves are the conic sections, the ellipse, parabola, and hyperbola, arising from the intersection of a plane and a cone. Intensively studied since ancient times, they apply to an enormous range of physical systems, from radar scanners to planetary orbits.

Curves in space

All the examples described so far are flat: the curves lie in a plane. But if both intersecting surfaces are nonlinear then the intersecting curve can twist around in space like a roller-coaster. Let's

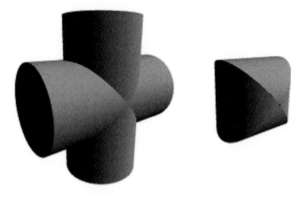

Fig 30.1 The intersection of two perpendicular cylinders consists of two ellipses (left). The volume of intersection is called a bicylinder (right).

consider the case of two cylinders, each with a circular cross-section, whose axes are at right angles and intersect in a point.

If the equations for the two cylinders are added, we obtain the equation of an *oblate spheroid*, a sphere flattened like an orange. If they are subtracted, we obtain an equation representing two planes. The actual intersection of the cylindrical surfaces comprises two ellipses. This may seem abstract, or even abstruse, but two perpendicular barrel vaults in a classical building intersect in this way, and we can clearly see the elliptical curves in the resulting groin vault (see Fig. 30.1).

The volume common to two cylinders of equal radii with orthogonal intersecting axes, called a *bicylinder*, was known to Archimedes, and also to the Chinese mathematician Tsu Ch'ung-Chih. In the fifth century, Tsu Ch'ung-Chih used it to calculate the volume of a sphere.

Now let's flatten the two cylinders so that they have elliptical cross-sections and displace them along the axis that is orthogonal to both of them, in opposite directions. Their equations are

$$2y^2 + (z + d)^2 = R^2, \qquad 2x^2 + (z - d)^2 = R^2,$$

where R is the radius of each cylinder and $2d$ is the separation between the two axes. Adding and subtracting these two equations, we get

$$x^2 + y^2 + z^2 = a^2, \qquad x^2 - y^2 = 2dz.$$

These are the equations for a sphere of radius a (where $a^2 = R^2 - d^2$) and another surface called a *hyperbolic paraboloid*. For brevity, let's call this a *hypar* (see Fig. 30.2).

In addition to its use in classical buildings, the hypar has proved useful in modern architecture. The advent of shell construction in the 20th century and the mathematical theory of surfaces allowed very thin, strong vaults to be constructed using the hypar form. Because it is a *ruled surface*, generated by straight lines, saddle-shaped roofs of this form are easily constructed from straight sections.

In Fig. 30.3, we plot the curve determined by the intersection of the sphere and hypar. It resembles the seam on a baseball or the groove on a tennis ball. The hyperbolic paraboloid is also the shape of the snack food called Pringles – the edge of a Pringle is like the tennis ball curve.

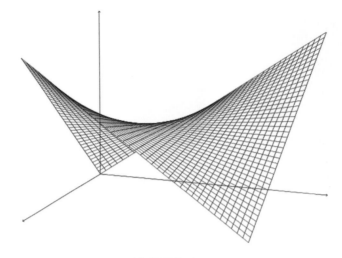

Fig 30.2 The hypar.

From tennis balls to weather forecasts

The groove on a tennis ball is not defined explicitly, and may be approximated in many ways. The official rules of the game are not much help, stating only that: 'The ball shall have a uniform outer surface of a fabric cover and shall be white or yellow in colour. If there are any seams, they shall be stitchless' (ITF Rules of Tennis 2012). The challenge is to construct a cover for the spherical ball from two flat pieces of felt. The great mathematician Carl Friedrich Gauss showed that to do this exactly is impossible: there is no exact mapping from a plane to a sphere. But, in practice, the felt flats are shaped like peanuts and, with a little stretching, fit snugly on the ball.

Many models of the tennis ball curve have been proposed. Indeed, the ingenious and versatile mathematician John Conway formulated a conjecture: No two definitions of 'the correct curve' will give the same answer unless their equivalence is obvious from the start. Put otherwise, there are numerous ways of defining the curve, all of which give similar but slightly different results. Indeed, the curve found on tennis balls is well approximated by a combination of four circular arcs. While this solution may appeal to engineers, it is unattractive to mathematicians, for the composite curve does not have nice analytical properties. The curve formed from the intersection of the sphere and hypar has an elegant mathematical equation. This curve is one of the numerous ways of defining the groove found on tennis balls.

The tennis ball curve arises from the practical need to cover the ball with flat felt. But the resulting partition of the sphere turns out to have another very practical use. In weather forecasting, we have to represent the atmosphere using a grid of points that cover the globe. The usual geographical latitude and longitude coordinates cause big problems: the meridians converge towards the poles, so the coverage with a latitude/longitude grid is highly non-uniform. By dividing the sphere into two parts by means of the tennis ball curve and using a separate grid on each part, we avoid the difficulties. This solution is called the Yin–Yang Grid, as it is reminiscent of the ancient Chinese symbol of that name.

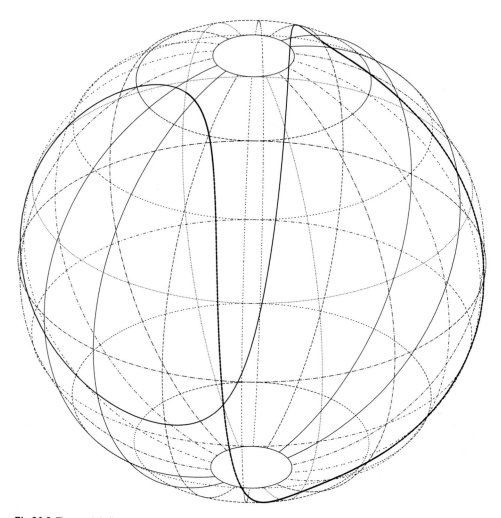

Fig 30.3 The tennis ball curve, the intersection of two offset elliptic cylinders, and also of a sphere and a hyperbolic paraboloid or hypar.

. . . .

FURTHER READING

[1] Tao Kiang (1972). An old Chinese way of finding the volume of a sphere. *The Mathematical Gazette*, vol. 56, pp. 88–90. Reprinted in *The changing shape of geometry*, edited by C. Pritchard, Cambridge University Press (2003).

[2] Robert Banks (1999). *Slicing pizzas, racing turtles, and further adventures in applied mathematics.* Princeton University Press.

CHAPTER 31

This is not a carrot: Paraconsistent mathematics

MAARTEN MCKUBRE-JORDENS

Paraconsistent mathematics is a theory of mathematics which makes sense of contradictions. In such a theory, it is perfectly possible for a statement *A* and its negation *not A* to both be true. How can this be, and be coherent? What does it all mean? Why, moreover, should we think mathematics might actually be paraconsistent? We will look at the last question first, starting with a quick trip into mathematical history.

Hilbert's programme and Gödel's theorem

In the early 20th century, the mathematician David Hilbert proposed that all of mathematics should be grounded on a small, elegant collection of self-evident truths, or *axioms*. Using the rules of logical inference, one should be able to prove all true statements in mathematics directly from these axioms. The resulting theory should be *sound* (only prove those statements that really are true), *consistent* (be free from contradictions), and *complete* (be able to either prove or disprove any statement).

However, the logician Kurt Gödel proved that this was impossible, at least in the sense that mathematicians of the time had in mind. His *incompleteness theorem*, loosely stated, says:

> In any formal theory that is free of contradictions and captures arithmetic, there are statements which cannot be proven true or false from within that theory.

As an example, consider a formal theory *T*: a system of mathematics based on a collection of axioms. Now consider the following statement *G*:

> G cannot be proved in the theory T.

If this statement is true, then there is at least one unprovable sentence in *T* (namely *G*), making *T* incomplete. On the other hand, if sentence *G* can be proved in *T*, we reach a contradiction: *G* is provable, but by virtue of its content, can also not be proven. Gödel showed that a sentence such as *G* can be created in any theory sufficiently sophisticated to perform arithmetic. Owing to this, mathematics must be either incomplete or inconsistent.

Classically minded scholars accept that mathematics must be incomplete, rather than inconsistent. They find contradictions abhorrent. However, accepting a small selection of contradictions need not commit you to a system full to the brim with them. Let us turn to a couple of cases where a paraconsistent position can provide a more elegant solution than the classical position: the paradoxes of Russell and the liar.

The liar paradox

For millennia, philosophers have contemplated the (in)famous liar paradox:

This statement is false.

To be true, the statement has to be false, and vice versa. Many brilliant minds have been afflicted with many agonising headaches over this problem, and there is not a single solution that is accepted by all. However, perhaps the best-known solution (at least among philosophers) is *Tarski's hierarchy*, named after the logician Alfred Tarski.

In a nutshell, Tarski's hierarchy assigns semantic concepts (such as truth and falsity) a level. To discuss whether a statement is true, one has to switch into a higher level of language. Instead of merely making a statement, one is making a statement about a statement. A language may only meaningfully talk about semantic concepts from a lower level. Thus a sentence such as the liar's sentence simply is not meaningful.

However, it intuitively *seems* that the liar sentence should be meaningful; it can be written down, it is grammatically correct, and the concepts within it are understood. It seems that to avoid inconsistency, classicists are forced to adopt some arguably ad hoc rules about the nature of meaning.

Russell's paradox

During his attempt to establish logical foundations for mathematics, Bertrand Russell discovered *Russell's paradox*. It concerns mathematical sets, which are, loosely speaking, just collections of objects. A set can contain other sets as its members: consider for example the set made up of the set of all triangles and the set of all squares. A set can even be a member of itself, for example the set T containing all things that are not triangles. Since T is not a triangle, it contains itself. Russell's paradox reads as follows:

Let R (the Russell set) be the set of all sets that are not members of themselves. Is R a member of R?

To be a member of itself, R is required not to be a member of itself. Thus if R is in R, then R is not in R, and vice versa.

Classical set theory was not equipped to deal with Russell's paradox. To avoid contradiction, mathematicians were forced to develop a whole new version of the theory, called *Zermelo–Fraenkel set theory* (ZF). In ZF, a hierarchical construction reminiscent of Tarski's ensures that Russell's paradox is avoided. This comes at a cost, however. It is difficult to motivate the array of different axioms on which ZF is built and they can be accused of being ad hoc. Moreover, ZF is an unwieldy system. Using a closely related system, Russell and Whitehead needed 379 pages to prove that $1 + 1 = 2$ in their *Principia Mathematica*, published in 1910.

Explosive logic

The paraconsistent response to these paradoxes is to say that they are interesting facts to study, instead of problems to solve. From a paraconsistent perspective, localised contradiction does not necessarily lead to global incoherence. How is this different from the classical view? For classicists, what is so bad about contradiction?

Classical mathematics uses classical logic to make inferences, and classical logic is *explosive*. An explosive logic maintains that from a contradiction, you may conclude anything and everything. If A and not A are both true, then Cleopatra is the current Secretary-General of the United Nations General Assembly, and the page you are currently reading is, despite appearances, also a carrot.

So why is classical logic explosive? It is because it accepts the argument form *reductio ad absurdum* (RAA), meaning reduction to the absurd. Essentially, the idea is that if assuming something is true leads to an 'absurd' state of affairs, a contradiction, then it was incorrect to make the initial assumption.

This seems to work well enough in everyday situations. However, if contradictions can be true, say if Russell's set both is and is not a member of itself, then we can deduce anything in a theory that admits the Russell set. We merely have to assume the negation of what we want to deduce, and then prove ourselves 'wrong' by noting the contradiction (this can be done formally using the rules of classical logic). So to the classical mathematician, finding a contradiction is not just unacceptable, it is utterly destructive. There is no classical distinction between inconsistency (the occurrence of a contradiction) and incoherence (a system which proves anything you like).

However, RAA raises the question of relevance. Suppose I have proved that the Russell set is and is not a member of itself. Why should it follow from this that there is a donkey braying loudly in my bedroom? This question has plagued classical logic for a long time.

Paraconsistent mathematics

Paraconsistency provides an alternative. It does not endorse the principle of explosion, nor anything that validates it. The thought is this: suppose I have a pretty good theory that makes sense of a lot of the things I see around me, and suppose that somewhere in the theory a contradiction is hiding. Paraconsistent logicians hold that this does not (necessarily) make the theory incoherent; it just means one has to be very careful in the deductions one makes. For the most part, it makes no difference to us if the liar sentence really is both true and false, and the paraconsistent perspective reflects that. *Reductio* proofs are no longer allowed in paraconsistent mathematics, since the conclusion could be a true contradiction, one which exists within the theory, and the logic must allow for this case. From the paraconsistent viewpoint, not all contradictions are necessarily absurd. Paraconsistentists can, however, salvage a form of RAA that allows them to reject something which is genuinely, paraconsistently absurd. This take on RAA is used to reject anything which leads to a trivial theory (a theory in which everything is true).

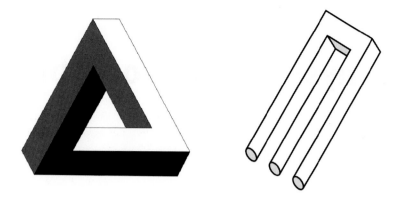

Fig 31.1 (Left) The Penrose triangle. (Right) The Blivet.

Allowing inconsistencies without incoherence opens up many areas of mathematics previously closed to mathematicians. One such area is inconsistent geometry. The *Penrose triangle* (Fig. 31.1, left) is a well-known example: its sides appear simultaneously to be perpendicular to each other and to form an equilateral triangle. The *Blivet* is another (Fig. 31.1, right), appearing to comprise two rectangular-box arms from one perspective, but three cylindrical arms from another. These pictures are inconsistent (as opposed to optical illusions as in Chapter 39), but at the same time coherent; certainly coherent enough to be put down on paper. Paraconsistent mathematics may allow us to understand these entities better.

Paraconsistency can also offer new insight into topics such as Gödel's incompleteness theorem. When Gödel tells us that mathematics must either be incomplete or inconsistent, paraconsistency makes the second option a genuine possibility. Classically, we assume the consistency of arithmetic and conclude that it must be incomplete. Under the paraconsistent viewpoint, it is entirely possible to find an inconsistent, coherent, and complete arithmetic. This could revive Hilbert's programme, the project of grounding mathematics in a finite set of axioms: if the requirement for consistency is lifted, it may be possible to find such a set.

Paraconsistency in mathematics: an interesting and promising position worthy of further exploration.

· ·

FURTHER READING

[1] Chris Mortensen (1996). *Inconsistent mathematics*. Stanford Encyclopedia of Philosophy. <http://plato.stanford.edu/entries/mathematics-inconsistent/> (site accessed May 2013).
[2] Graham Priest (2006). *In contradiction*. Oxford University Press.
[3] Zach Weber (2009). *Inconsistent mathematics*. Internet Encyclopedia of Philosophy. <http://www.iep.utm.edu/math-inc/> (site accessed May 2013).

Source
An earlier version of this article appeared in *Plus Magazine* (<http://plus.maths.org>) on 24 August 2011.

CHAPTER 32

The mystery of Groombridge Place

ALEXANDER MASTERS AND SIMON NORTON

'It is a murder with mathematical significance,' grunted Simon with satisfaction. 'He was discovered with his face blown off by a shotgun.'

'And his widow was suspiciously jolly in the box hedge garden.'

'Sherlock Holmes's arch-enemy, Professor Moriarty, behind it all,' agreed Simon, pleasurably.

For those of you who don't know about Simon Phillips Norton, a few words of introduction are needed. A child prodigy, in the 1960s Simon won the top award at the International Mathematics Olympiads three times, twice scoring 100%. He made headlines in the *Daily Mail*, the *Telegraph*, and the *Daily Sketch*. When he arrived at Trinity College, Cambridge, aged 17, Simon had already secured a First Class degree in Pure Mathematics at London University.

Love (for public transport, not ladies) put an end to Simon's career. These days, Simon spends between four and seven days every week jaunting around the country on bus and train trips, admiring the views and complaining to local authorities about inefficient connections. The obsessional energy of his early genius has been redirected. He hasn't given up mathematics; he is still a genius. When he can be bothered, his contributions at international group theory conferences are always anticipated with excitement; but mathematics is no longer the centre of his life. His grey hair is electrified; his nylon trousers, often torn; his T-shirt is speckled with stains from his favourite dish of mackerel and Chinese-flavoured packet rice. To non-mathematicians like myself and to mathematicians terrified of losing their own talent, incapable of leaving with grace the excitement of being the best in the room, Simon's move away from top-level professional mathematics might seem to represent a devastating failure. He had been the greatest and has become ... a bus and train spotter. But to Simon, it has been a blossoming. He edits an entertaining campaign newspaper (<http://www.cambsbettertransport.org.uk>) dedicated to the end of private motor cars and the development of a well-integrated public transport service. Never lonely or bored, always occupied, fighting a great and unselfish cause in order to make the world a better place, his move away from mathematics represents a triumph of self-awareness and happiness. Apart from public transport timetables, Simon has one other passion: murder mysteries.

That's why we were in Tunbridge Wells West a few weeks ago, sitting on the 10.50 steam train service to Groombridge, recollecting delicious details from *The valley of fear*. Groombridge Place was Conan Doyle's model for Birlstone Manor, the scene of the crime.

Simon plunged both his arms into the holdall he had put beside him, and began beating his hands around among the leaflets and emergency tins of mackerel he keeps in there. 'I have developed a new problem which is relevant to the matter of this murder,' he called, his mouth pressed against his chest as he peered into the bag's darkness. 'Oh dear, oh dear!' he sighed, unable to locate the paper he'd written on the subject. But a minute later he pulled out a besplodged pair of typed pages in a minuscule font, and handed them over with triumph. 'It involves calculus, trigonometry, complex numbers, and the regraduation of time.'

Our train gave a piercing toot and set off for Groombridge. Outside, smoke from the engine tumbled down the platform, swirled among the snowflakes, and poured over the station picket fence into the ASDA car park. 'And it explains why Moriarty became the Napoleon of Crime,' Simon added with glee. The distance from Tunbridge Wells to Groombridge is very small, and in order to justify the £7 fare the train has to proceed extremely slowly. As the engine banged and grumbled along the rails, this is what I read:

One of the joys of reading fiction is speculating on how the world created by the author can relate to the real world. No series has been more successful in this respect than Conan Doyle's Sherlock Holmes stories, which have spawned a considerable literature of interpretation based on the pretence that Sherlock Holmes was a real person.

Mathematics comes into the interpretation of the Sherlock Holmes stories because his arch-enemy, Moriarty, was at one time a mathematics professor. Two of his publications are mentioned: a treatise on the binomial theorem in *The final problem* and an article, '*The dynamics of an asteroid*', in *The valley of fear*. One might speculate what these papers were actually about.

Back in the real world, in 1887 King Oscar II of Sweden offered a prize for a solution to the problem of describing the motion of a multiplicity of bodies such as the solar system under the action of Newtonian gravity. For two bodies, the answer to the problem was known to Newton, who showed that his gravitational theory implied as a consequence that they would orbit around their centre of gravity in elliptic paths as per the laws already established by the observations of Kepler. However, for as few as three bodies, the problem becomes mathematically intractable. For many asteroids, the only significant gravitational forces they feel are those exerted by the Sun and Jupiter, so a description of their motion would form part of the three-body problem.

The prize was won by the French mathematician Henri Poincaré; he didn't actually solve the problem, but he made the first major inroads into it.

Returning to the world of Sherlock Holmes, let us, therefore, speculate that *Dynamics of an asteroid* was Moriarty's entry for the prize. Let us further speculate that King Oscar's advisers became suspicious of this entry, and commissioned Holmes to investigate its authenticity. *The final problem* includes a reference to work by Sherlock Holmes on behalf of 'the King of Scandinavia', which sounds like a reference to Oscar.

If Holmes found evidence that the work was plagiarised, then Moriarty would have come under pressure to resign his chair – as we're told he was forced to do in *The final problem*. Could his grievance against Holmes have led him to take up a new career in organised crime as a kind of vendetta? Perhaps, in some kind of deal to minimise the scandal, Holmes would have agreed not to publicise the plagiarism, which would explain why in *The valley of fear* he refers to *Dynamics of an asteroid* as a major achievement.

Larry Millett, of St Paul, Minnesota, has written several books about Sherlock Holmes's adventures in America. In one of them, *Sherlock Holmes and the rune stone mystery*, Holmes is commissioned by the very same King Oscar to investigate the authenticity of runic inscriptions found on a stone near the Minnesotan town of Alexandria. This stone is modelled on one that actually exists and is on display in the museum at Alexandria. It was, in fact, reading this book that led me to speculate the following: if Holmes had successfully resolved the authenticity of *Dynamics of an asteroid*, wouldn't it be more likely that King Oscar would use him again to test the authenticity of the rune stone?

Perhaps some writer will come up with a book called *Sherlock Holmes and the three-body problem* about Holmes's first service for King Oscar, where the title would have a double meaning in that it also refers to three human bodies who are murdered in the course of the plot. If so I'd love to read it!

In the meantime, here is a three-body problem that Moriarty would have had no difficulty with:

Three-insect problem: Three insects, A, B, and C, which should be considered as points, lie on a plane surface. A moves towards B, B towards C, and C towards A [see Fig. 1], each changing its direction so that it moves straight towards its target, and each moving at the same speed. Describe their motion.

This differs from the gravitational three-body problem in several ways. First of all, the gravitational problem is *chaotic*, that is, one may well need to know the starting position with infinite precision to describe the motion, while the three-insect problem is *regular*, so that the motion can be described in a fairly uniform manner. Secondly, there are far fewer degrees of freedom in the three-insect problem. If one knows the positions of the insects at time zero, then everything else is determined. Each position has two coordinates. Given that the system is invariant under rotations, reflections, and so on (the problem remains the same when you apply those transformations to the plane), there are in fact only two degrees of freedom left.

By contrast, for the gravitational problem, one needs to know the initial speed of each object as well as its position, and that makes up no less than twelve degrees of freedom (if one assumes that the centre of gravity is fixed at a given place, as one can do without loss of generality), of which seven can be removed by applying congruences and similarities, leaving five. (If the three bodies are assumed to stay in the plane on which they lie, there are still four degrees of freedom.)

In both problems there are *easy cases*. For the gravitational problem, these are when a body is at its so-called Lagrange points: if one considers the three bodies as being the Sun, Jupiter, and an asteroid, then the Lagrange points

Fig 1 The three-insect problem, showing the instantaneous directions of motion of each insect.

for the asteroid are the two points which form an equilateral triangle in the plane of Jupiter's orbit with the Sun and Jupiter, and three points on the line joining Jupiter with the Sun (one each with the Sun, Jupiter, and the asteroid between the other two).

Curiously, the easy cases for the three-insect problem are very similar: if the insects form an equilateral triangle they will all converge on the triangle's centre along logarithmic spirals (this is a case which is often set as a puzzle), while if they lie on a line then the insects at each end will travel straight towards each other while the other one will travel towards its target and then stick to it.

There is, however, an interesting difference between the two problems here: for the gravitational problem, the triangular cases are stable in that if the bodies start close to this configuration they will converge on it, while the straight-line cases are unstable in that if they start close to them they will move further away; but in the three-insect problem it is the equilateral triangle that is unstable, while the straight-line cases are stable.

Here is an outline of the mathematics involved in the three-insect problem, which has some elegance. Consider the triangle formed by the insects. We have already given the insects the labels A, B, and C; let us use the same labels for the angles in the triangle they form, and the labels a, b, and c for the respectively opposite sides. This conforms with the standard labelling convention for triangles in geometry.

Now B is moving towards C at unit speed. But C's motion will also be affecting its distance from B: if the angle C is acute then C will be moving towards B and if the angle C is obtuse then C will be moving away. In both cases (and also when the angle is a right angle) the rate at which C is moving towards B is $\cos C$: if the angle is obtuse this is negative, which means that C is moving towards B at a negative rate, that is, moving away.

As B is moving towards C with speed 1 and C is moving towards B with speed $\cos C$ it follows that

$$-\frac{da}{dt} = 1 + \cos C \quad \text{and, similarly,} \quad -\frac{db}{dt} = 1 + \cos A \quad \text{and} \quad -\frac{dc}{dt} = 1 + \cos B.$$

This means that over time the perimeter of the triangle, $a + b + c$, is decreasing at the rate $3 + \cos A + \cos B + \cos C$. It can be shown that whatever the shape of the triangle this lies between 4 (if the insects are in a line) and $4\frac{1}{2}$ (if they form an equilateral triangle). Thus we know to within about 6% how long the insects will take to converge on a point.

Now we apply the familiar cosine rule

$$c^2 = a^2 + b^2 - 2ab \cos C,$$

and the similar equations obtained by permuting the letters a, b, and c (in both lower and upper cases). The first of the above equations then becomes

$$-\frac{da}{dt} = 1 + \frac{a^2 + b^2 - c^2}{2ab} = \frac{(a + b + c)(a + b - c)}{2ab}.$$

These equations can be simplified if we regraduate time by multiplying each of da/dt, db/dt, and dc/dt by $2abc/(a + b + c)$ and introducing a new set of parameters, $x = b + c - a$, $y = c + a - b$, $x = a + b - c$. In terms of these new variables we have

$$-\frac{dx}{dt} = xy, \quad -\frac{dy}{dt} = yz, \quad \text{and} \quad -\frac{dz}{dt} = zx.$$

These equations are now much simpler. Note that x, y, and z are always positive (or possibly zero if the insects are in a straight line), because of the so-called triangle inequality, which states that the sum of two sides of a triangle is always greater than the third.

But at that point the engine tooted, the points rattled, and we clanked into Groombridge station in a swirl of snow. As we walked across the muddy fields to the scene of the crime, Simon was unable to hold back his enthusiasm for his performing insects. One of the joys of knowing Simon is that he is always delighted to try to explain mathematics, even at a murder scene in a blizzard. 'We now encode the relative positions of the insects by a point inside an equilateral triangle Δ chosen so that its distances from the three sides are proportional to x, y, and z. This means that any two configurations of insects that correspond to the same point are congruent or similar, and will evolve in exactly the same way. Then the following can be shown:

'(a) The path of the point always rotates the same way round the centre of Δ (that is, either clockwise or anticlockwise depending on how we draw Δ);

'(b) $x/y + y/z + z/x$ is continually increasing.'

'As a reader of popular mathematics, I find it very irritating when I read an article that talks about a problem but doesn't give the solution, even though the solution is well within the capability of someone with a layman's knowledge of mathematics.'

In the gardens of the house we sat on a bench that looked across the moat, into the room where the victim's head had been blown off

... we now conclude that, unless the insects start out in an equilateral triangle or all in a line, the triangle formed by the three insects tends to flatten out so that they all lie close to a line. The motion on the flattened triangle, though, resembles that of an approximate spiral. So, each of the insects takes turns in becoming the one in the middle.

This seems to give a pretty good idea of what's going on, though there are still a number of questions one might want to ask which we don't have the answers for.

and I glanced at Simon's pages again. Here he'd written out the conclusion in simpler language.

'To put it into context,' said Simon, 'after I thought of the problem it didn't take me all that long to get it into the form shown in the equations I have just given you, but the problem was simmering for a long time before I was able to work out how to complete the solution.'

For all his generosity and eagerness to help, he can't see mathematics the way ordinary people do. For him, it's not something you spend years agonising over, until you can finally stretch up and grab a few rewards; it's more like a cave filled with diamonds. To Simon's fertile mind, all 'the layman' has to do is learn the few elementary tricks needed to break into the cave, after which life sparkles with mathematical delights forever. Why struggle? Why agonise? Just wander about the cave and enjoy. The gems are there for the taking.

The rest of the afternoon, Simon was half in the gardens of Groombridge House, half in his cave of glittering things. 'With some further work, those results (a) and (b) can be shown to imply that the point follows a spiral path which comes closer and closer to the three sides of Δ. This means that the insects come closer and closer to a straight line, with each taking it in its turn to be the one in the middle.'

He paused and sank his hands into his holdall to chase out a bag of Bombay Mix. Then, abruptly, he delivered his astronomical conclusion:

'This kind of "choreography", or periodic exchange of roles, is a bit unexpected, but it has also been seen in some cases of the gravitational three-body problem.'

For a few minutes there was only the sound of the crackling Bombay Mix bag and Simon's chewing.

'My full paper about the problem can be read in *Mathematics Today*,' he said and yawned enormously, startling a passing peacock. He returned his snack to the holdall and stood up. 'The readers will understand that a little better than you have, perhaps, but I think even you must agree that it sheds light on the relationship between Moriarty and Sherlock Holmes.' As Watson said in 'The problem of Thor Bridge' (*The Casebook of Sherlock Holmes*): 'A problem without a solution may interest the student, but cannot fail to annoy the casual reader'.

PYTHAGORAS'S THEOREM: b^2
Hommage à Queneau

Pythagoras deconstructed (extract)

It should be clear that the Pythagorean Mystery generates a number of epistemological tensions. Had it been a left-angled triangle then the effects might have been somewhat mitigated: the left is, after all, *sinister* (from the Latin), or *left over* (unnecessary, prone to be thrown out), or *left off* (deleted, censored) or most powerfully *left alone* (isolated, outcast). The right-angled triangle should be upstanding, correct, one is tempted to say legal or straight (*droit*), and hence the natural/unnatural contradiction of the appearance of irrationality in this context can only serve to exacerbate the discomfort of the reader in his or her enforced participation ...

Proof by experimentalist

$$0.97c^2 < a^2 + b^2 < 1.01c^2.$$

P.G. Wodehouse

'I say, Jeeves, what the bally ho is a hypotenuse?'
'A hypotenuse is the longest side of a right-angled triangle, Sir.'
'Sounds a bit square to me, Jeeves.'
'More than a bit, Sir, in fact as square as the other two sides put together.'
'Hot dog, Jeeves, it all adds up!'
'Certainly, Sir!'

Mr Chips

TEACHER: The interesting point here is that for a right-angled triangle the *Riemann, will you stop bending your exercise book like that, you'll end up by breaking it* – the square on the hypotenuse, that's the long side, *is what is it, Deligne? No. You cannot reverse the triangle inequality, that would make no sense at all, you stupid boy. Where was I? Ah,* the square on the hypotenuse equals the sum of the squares *I'm watching you, Pierre, I said SQUARES* sum of the squares of the other two sides. The proof due to Euclid uses *don't get smart with me, Isaac, it was good enough for Euclid and I don't care if you think you have a simpler proof – Stop scribbling in the* margin, *Pierre Euclid's proof uses congruent Albert, how many times do I have to tell you that staring at sunbeams like that will make you blind?. . . .*

London Underground

Fairlop's Last Theorem:

* Leicester Square via Russell Square equals Euston Square. Mornington Crescent!

[Editorial note: The above play assumes the first Pythagorean convention, in which the use of diagonals is permitted.]

The instruction manual

1 Place all the pieces of your TRIANG®-L on a flat surface, making sure to dispose of packaging responsibly.

2 Locate the two acute angle-joiners α and β. Attach to each end of side H.

3 Take side O and join to free socket at β taking care to ensure that both O and H are free to rotate.

4 Similarly attach side A to α.

5 Taking the free ends of A and O (while ensuring H remains fixed), rotate so that they are just touching.

6 Use the ANGLETESTER® provided to measure the angle between A and B. If correctly aligned, this should read 90°. You may need to use a knife to trim the ends of A or O.

7 Now use the supplied RIGHTANGLE® R to join A and B. If correctly aligned, there should be no overlap.

8 You are now free to enjoy TRIANG®-L. Please use with care.

Estate agent

The Triangle benefits from generous proportions, its sweeping hypotenuse having a glorious southerly view over the plane. Some elements of this well-appointed property can be traced back to ancient Greece – it is rumoured Pythagoras once rested his pencil in the drawing room. '*a*' magnificent '*c*' view can '*b*' glimpsed from the right angle through acute window in the northeastern aspect.

William Blake

Tryangle! Tryangle! angle ninety
In the mathbook of the mighty,
What immortal hand or eye
Could frame thy fearful symmetry?

And what short sides left and right
Could sum to make the longer flight?
How to add them to make compare
The hypotenuse? raise to square!

e.e. pythagoras

Look at the triangle in your head
it has an

```
                      i
                   m
                 a
               g
            i
          n
        a
      r
    y
```

hypotenuse
and if you sit
    in the sun
       on the square
          of the o
                t
                h
                e
                r
        s e d i s   o w t
it fits perfectly
and all is quiet again

Limerick

A right-angled triangle opined
My hypotenuse squared is refined
For if anyone cares
It's the sum of the squares
Of my other two sides when combined.

Tabloid newspaper story

A Greek maths boffin has found the secret formula for summing sides of triangles.

The bust Greek banks mean that beaches and tavernas in that part of the Med are now as cheap as chips (see Reader Offer on Page 52). But it's not all sea, sex, and sangria.

Bored of bonking

The hunky maths prof, 36, with the unlikely name of Mr Pythagoras has come up with a new pastime for those bored of bonking.

It is said his inspiration came from the love triangles he sees every summer as the tanned holidaymaking Brits come back from the beaches.

'Imagine,' says the boff, 'your beer is on one side of the bar while your bird is on the other. What do you do?' His formula has the answer.

Your reporter got bored during the tekky bits, but the magic formula apparently involves being square and having sideburns (yuk!).

New sport of math making

Spokeswoman for the Greek Tourist Office, the buxom 26 year old Titania Petrapopytopoulous – vital statistics 38-25-33 – claims the breakthrough could boost the flow of sunseekers to holiday resorts.

'Maths is the new sport over here' she claims. 'What better after a day on the beach and a night on the tiles, than to add, subtract, or multiply?'

It all sounds Greek to us. Not sure it will catch on back home.

CHAPTER 33

Mathematicians at the movies: Sherlock Holmes vs Professor Moriarty

DEREK E. MOULTON AND ALAIN GORIELY

7,2–9,2–6,0 15,0–10,2–5,0 10,9–10,1–9,3_5,0–7,3–7,2 7,2–3,0–6,2 13,2–8,1–4,0_12,18–6,0–1,0 15,4–7,0–5,0 6,0–5,0–6,1*

'The world is full of obvious things which nobody by any chance ever observes.'

Sherlock Holmes

During the summer of 2010, OCCAM (the Oxford Centre for Collaborative Applied Mathematics) received a call from Warner Bros. The mission, should we choose to accept it, was to help Warner Bros with the mathematical aspects of the new movie *Sherlock Holmes: A Game of Shadows*, in which Sherlock's arch-enemy is the mathematician Professor James Moriarty. Our initial task was to design the equations appearing on the giant board in Moriarty's office. They were to be historically accurate (around 1890) and potentially reveal some of Moriarty's evil plans. The initial request soon grew from simply designing equations to devising a secret code, creating Moriarty's lecture while on his European tour, and providing suggestions on how Moriarty would use mathematics to carry out his plots and how Sherlock, in turn, deciphers them. Not surprisingly, very little mathematics actually made it to the big screen. Nevertheless, interesting snippets did make the final cut if you know where to look. In particular, a close examination of Moriarty's giant board tells the whole mathematical story, and we decipher it for you in this article.

Arthur Conan Doyle actually gave little information on Moriarty, apart from being described by Holmes as 'a mathematical genius' and 'the Napoleon of crime'. On the academic side, we know that he wrote a treatise on the binomial theorem and that 'On the strength of it, he won the mathematical chair at one of our smaller universities' (Holmes, in *The final problem*). Moriarty is also said to have written *The dynamics of an asteroid*, which Sherlock describes as 'a book which ascends to such rarefied heights of pure mathematics that it is said that there was no man in the scientific press capable of criticising it'. Taking these cues, we delved into the mathematical mind of Moriarty and the fascinating mathematics at the turn of the 20th century.

---

*Contact the authors if you crack the code (they can be found on the web); a prize awaits the first to do so.

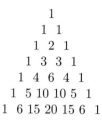

```
              1
            1   1
          1   2   1
        1   3   3   1
      1   4   6   4   1
    1   5  10  10   5   1
  1   6  15  20  15   6   1
```

Fig 33.1 Pascal's triangle.

The code

One of the primary elements on the board, and a key aspect in Moriarty's evil plans, is his secret code that he uses to communicate with his henchmen as well as encrypt his own information about his global empire. Given his obsession with the binomial theorem, we created a code based on Pascal's triangle (see Fig. 33.1). The code has three elements: a public key, a coded formula, and a cipher. To code information, Moriarty first pieces together his message from words taken from different locations in a horticultural book that he keeps in his office (this book does not actually exist; a modern version of the cipher would rely on a classic book). Each word of the message corresponds to three numbers: the page, line, and word numbers. Through this process, the actual message is converted into a structured sequence: the *book sequence*. Moriarty takes the book sequence, and encodes it further using Pascal's triangle.

The message requires a public key, a whole number p. From each number p, one can build a sequence of numbers, the Fibonacci p numbers, denoted by F_p. The sequence is defined by $F_p(n) = F_p(n-1) + F_p(n-p-1)$ with $F_p(1) = 1$ and $F_p(n) = 1$ for $1 - p \leq n \leq 0$ and can be created by summing along the pth diagonal of Pascal's triangle. For $p = 0$, we recover the powers of 2 (the sum of the horizontal lines in the Pascal triangle), whereas $p = 1$ corresponds to the classical Fibonacci sequence (sum of the first diagonal in the Pascal triangle).

Having chosen p, any integer N may be represented by giving the minimal representation: the unique representation $N = F_p(n) + \phi$, with $\phi < F_p(n - p)$. In this way, N is encoded as n, ϕ, and Moriarty thus converts the book sequence into a new, fully coded sequence.

In coding his messages to his associates, Moriarty must pass them the public key so that they know which Fibonacci numbers to use in decoding the message. This number is included at a key moment in Moriarty's lecture, by changing the value of a particular variable.

Cracking the code

Holmes observes work related to Pascal's triangle and Fibonacci p numbers written on Moriarty's board in his office. Later, he notices a slight difference in Moriarty's lectures, guiding him to the idea that an integer key is being passed to an associate, which he eventually realises corresponds to the particular p in the Fibonacci p numbers. Holmes deduces that the horticultural book he saw in Moriarty's office is serving as a cipher based on the fact that the flower in Moriarty's office is dying. His powerful intellect does the rest. We in fact suggested, with no luck, that the dying flower should naturally be a sunflower head, which would have given another nice connection between Fibonacci numbers and phyllotaxis.

The lecture

The other key mathematical element in the story is the lecture tour Moriarty gives. The goal was to design a plausible lecture that Moriarty could have given around 1895 that fits with his work on the dynamics of asteroids, and that was important enough to warrant a lecture tour around Europe. While on the surface the lecture tour is a vehicle for Moriarty to oversee and orchestrate his evil empire, we decided that the lecture itself should be explosive and characterise the villain and his motivation for crime.

We turned to two major works of celestial mechanics around the end of the 19th century. The first one is the work of Henri Poincaré on the so-called n-body problem (initially known as the *three-body problem*). The n-body problem consists in finding solutions of Newton's equation of universal gravitation for the interaction of n masses: how will the gravitational pull that these masses (for example planets) exert on each other affect their movement over time? The problem was so important for the time that Oscar II, King of Sweden and Norway, announced a special prize to the mathematician who could solve the problem. The prize winner was Henri Poincaré. His work on the topic was expanded in a series of three books (*New methods of celestial mechanics*) published in 1892. These books are regarded as some the most influential works of the early 20th century and are usually attributed as the origin of important concepts of mathematics such as geometric analysis for dynamical systems, chaos, and asymptotic expansions.

A second work of great importance for Moriarty's lecture is the work of Paul Painlevé. In 1895, Painlevé was invited to Stockholm (again by King Oscar II) to deliver a series of lectures on his work. The event was considered so important that King Oscar himself attended the opening lecture. Of relevance for Moriarty's lecture is Painlevé's work on collisions. Painlevé looked at the possibility of collisions between masses in gravitational interactions and proved some fundamental results complementing the work of Poincaré.

In particular, he discussed whether solutions to Newton's equation could become infinite at a finite time (these are called *finite-time singularities*). As an example, he considered the system

$$\frac{d^2x}{dt^2} = \frac{(y-x)}{(x^2+y^2)}, \quad \frac{d^2y}{dt^2} = \frac{(-y-x)}{(x^2+y^2)}, \quad \frac{d^2z}{dt^2} = \frac{(x-py)}{(x^2+y^2)^2}.$$

For the particular case $p = 3$, the system has a finite-time singularity at t_* given by $x = (t_* - t) \sin[\log(t_* - t)]$, $y = (t_* - t) \cos[\log(t_* - t)]$, $z = (t_* - t)^{-1} \sin[\log(t_* - t)]$ (the phase space of which is sketched on Moriarty's board – this is the plot of velocity versus displacement). Painlevé used this example to motivate his discussion and proof of the non-existence of singular solutions for the three-body problem, in the case where none of the bodies collide with each other. Also of historical importance is the fact that Painlevé would become the Minister of War during the Great War, another interesting connection with Moriarty's obsession with weaponry. In many ways, Painlevé is the man that Moriarty could have become if he had been properly recognised and did not develop his 'hereditary tendencies of the most diabolical kind'.

With the villain and his mathematics firmly established in our minds, we produced pages of possible lecture material, secret codes, and evil plots, complete with Holmes' logical deductions to unravel it all.

Epilogue

In December 2010 we were invited on set at Hatfield House near London for the filming of the office scene, where, as far as we were concerned, the central character was to be the blackboard

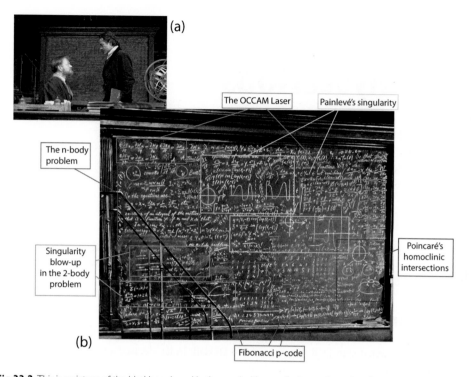

(a)

The OCCAM Laser

Painlevé's singularity

The n-body problem

Singularity blow-up in the 2-body problem

Poincaré's homoclinic intersections

(b)

Fibonacci p-code

Fig 33.2 This is a picture of the blackboard used in the movie. You can find a number of mathematical elements discussed in the text. You can also spot four different instances where the word OCCAM is spelled out. It appears that this acronym may have found its way into the movie accidentally.

(Fig. 33.2). The day was as cold and miserable as only English winter days can be. After waiting patiently for many hours, we finally got to see the board. It was beautiful, imposing, and ... full of typos. We spent countless hours helping the professional calligrapher rectify the mistakes on the board and teaching the crew about the subtleties of subscripts and curly derivatives. But, by the end of the day, two simple facts of life became obvious to us: first, in the totem pole of Hollywood, mathematicians sit comfortably at the very bottom, and second, the long and uneventful hours of waiting and preparation made us realise how charming and pleasant our office life truly is: there is really no business like academic business.

A year later, the movie came out and we were invited to the screening in London. The movie is rather long and violent (a fast-paced action movie, they would call it in Hollywood). Clearly, this new Sherlock Holmes is more muscular and physical than any others before. His eventual victory is as much based on his usual deductive powers as on an uncanny ability to predict un-reasonably well the combined outcome of many random events, all in slow motion. This is rather unfortunate; Sherlock's timeless, enduring quality is not an unphysical skill of predicting where his opponent will punch 20 moves ahead. No, his real skills are his ability to use his broad sci-entific knowledge, his logical mind, and his perseverance to piece together unrelated information into a single unified picture in order to crack intellectual puzzles. Indeed, Sherlock would have made a fine applied mathematician.

Source

A different version of this article originally appeared in *SIAM News*, Vol. 45(3), April 2012, pp. 1–2.

CHAPTER 34

The mysterious number 6174

YUTAKA NISHIYAMA

The number 6174 is a really mysterious number. At first glance, it might not seem so obvious. But as we are about to see, anyone who can subtract can uncover the mystery that makes 6174 so special.

Kaprekar's operation

In 1949 the mathematician D. R. Kaprekar from Devlali, India, devised a process now known as *Kaprekar's operation*. First choose a four-digit number in which the digits are not all the same (i.e. not 0000, 1111, 2222, . . .). Then rearrange the digits to get the largest and smallest numbers these digits can make. Finally, subtract the smallest number from the largest to get a new number, and carry on repeating the operation for each new number.

It is a simple operation, but Kaprekar discovered it led to a surprising result. Let's try it out, starting with the number 2014, the digits of the IMA's 50th anniversary year. The maximum number we can make with these digits is 4210, and the minimum is 0124 or 124 (if one or more of the digits is zero, embed these in the left-hand side of the minimum number). Iterating this procedure gives

$$4210 - 0124 = 4086,$$
$$8640 - 0468 = 8172,$$
$$8721 - 1278 = 7443,$$
$$7443 - 3447 = 3996,$$
$$9963 - 3699 = 6264,$$
$$6642 - 2466 = 4176,$$
$$7641 - 1467 = 6174.$$

When we reach 6174 the operation repeats itself, returning 6174 every time. We call the number 6174 a *kernel* of this operation, and in fact it is the only one. But 6174 also has one more surprise up its sleeve. Let's try again starting with a different number, say the revolutionary 1789:

$$9871 - 1789 = 8082,$$
$$8820 - 0288 = 8532,$$
$$8532 - 2358 = 6174.$$

We've reached 6174 again!

When we started with 2014 the process reached 6174 in seven steps, and for 1789 it took three steps. In fact, you reach 6174 for all four-digit numbers that don't have all the digits the same. It's marvellous, isn't it? Kaprekar's operation is so simple but uncovers such an interesting result. And this will become even more intriguing when we think about the reason why almost all four-digit numbers reach this mysterious number 6174.

Is it only the number 6174 that has this property?

The digits of any four-digit number can be arranged into a maximum number by putting the digits in descending order, and a minimum number by putting them in ascending order. So for four digits a, b, c, d, where

$$9 \geq a \geq b \geq c \geq d \geq 0$$

and a, b, c, and d are not all the same digit, the maximum number is $abcd$ and the minimum is $dcba$. We can calculate the result of Kaprekar's operation using the standard method of subtraction applied to each column of this problem:

$$abcd - dcba = ABCD.$$

If none of the digits are equal, so $a > b > c > d$, this implies that

$$
\begin{aligned}
D &= 10 + d - a & \text{(as } a > d), \\
C &= 10 + c - 1 - b = 9 + c - b & \text{(as } b > c - 1), \\
B &= b - 1 - c & \text{(as } b > c), \\
A &= a - d.
\end{aligned}
$$

A number will be repeated under Kaprekar's operation if the resulting number $ABCD$ can be written using the initial four (distinct) digits a, b, c, and d. So we can find the kernels of Kaprekar's operation by considering all the $4 \times 3 \times 2 = 24$ combinations of $\{A, B, C, D\}$ in terms of $\{a, b, c, d\}$ and checking if they satisfy the relations above. Each choice gives a system of four simultaneous equations with four unknowns to solve for a, b, c, and d.

It turns out that only one of these combinations has integer solutions which satisfy $9 \geq a > b > c > d \geq 0$. That combination is $ABCD = bdac$, and the solution to the simultaneous equations is $a = 7$, $b = 6$, $c = 4$, and $d = 1$, i.e. $abcd = 6174$. There are no valid non-zero solutions to the simultaneous equations resulting from some of the digits in $\{a, b, c, d\}$ being equal. Therefore the number 6174 is the only non-zero number unchanged by Kaprekar's operation – our mysterious number is unique.

For three-digit numbers the same phenomenon occurs. For example, applying Kaprekar's operation to the three-digit number 753 gives the following:

$$
\begin{aligned}
753 - 357 &= 396, \\
963 - 369 &= 594, \\
954 - 459 &= 495, \\
954 - 459 &= 495.
\end{aligned}
$$

The number 495 is the unique kernel for the operation on three-digit numbers, and all three-digit numbers reach 495 using the operation. Why don't you check it yourself?

Two digits, five digits, six, and beyond . . .

We have seen that four- and three-digit numbers reach a unique kernel, but how about other numbers? It turns out that the answers for those are not quite as impressive. Let's try it out for a two-digit number, say 28:

$$82 - 28 = 54,$$
$$54 - 45 = 9,$$
$$90 - 09 = 81,$$
$$81 - 18 = 63,$$
$$63 - 36 = 27,$$
$$72 - 27 = 45,$$
$$54 - 45 = 9.$$

It doesn't take long to check that all two-digit numbers will reach the loop $9 \rightarrow 81 \rightarrow 63 \rightarrow 27 \rightarrow 45 \rightarrow 9$.

Unlike for three- and four-digit numbers, there is no unique kernel for two-digit numbers.

But what about five digits? Is there a kernel for five digit numbers like 6174 and 495? To answer this we would need to use a similar process as before: check the 120 combinations of $\{a, b, c, d, e\}$ for $ABCDE$ such that

$$9 \geq a \geq b \geq c \geq d \geq e \geq 0$$

and

$$abcde - edcba = ABCDE.$$

Thankfully the calculations have already been done by a computer, and it is known that there is no kernel for Kaprekar's operation on five digit numbers. But all five digit numbers do reach one of the following three loops:

$$71973 \rightarrow 83952 \rightarrow 74943 \rightarrow 62964 \rightarrow 71973$$
$$75933 \rightarrow 63954 \rightarrow 61974 \rightarrow 82962 \rightarrow 75933$$
$$59994 \rightarrow 53955 \rightarrow 59994.$$

As Malcolm Lines points out in his book, *A number for your thoughts*, it takes a great deal of time to check to see what happens for six or more digits, and this work becomes extremely dull! To save you, my number-experimenting reader, from this fate, Table 34.1 shows the kernels for two-digit to ten-digit numbers. It appears that Kaprekar's operation takes every number to a unique kernel only for three- and four-digit numbers.

Kaprekar's operation is so beautiful we may feel that there must be something more to its beauty than just chance. Is it enough to know that almost all four-digit numbers reach 6174 by Kaprekar's operation, but not know the reason why? So far, nobody has been able to show that the fact that all numbers reach a unique kernel for three- and four-digit numbers is anything more than an accidental phenomenon. This property seems so surprising it leads us to think that maybe, just maybe, an important theorem in number theory is hiding in Kaprekar's numbers waiting for a mathematician of the future to discover it.

Table 34.1 Kernels for 2- to 10-digit numbers

Digits	Kernel
2	None
3	495
4	6174
5	None
6	549945, 631764
7	None
8	63317664, 97508421
9	554999445, 864197532
10	6333176664, 9753086421, 9975084201

· ·

FURTHER READING

[1] Martin Gardner (1985). *The magic numbers of Doctor Matrix*. Prometheus Books.
[2] D. R. Kaprekar (1949). Another solitaire game. *Scripta Mathematica*, vol. 15, pp. 244–245.
[3] Malcolm Lines (1986). *A number for your thoughts: Facts and speculations about numbers from Euclid to the latest computers*. Taylor and Francis.
[4] Yutaka Nishiyama (2013). *The mysterious number 6174: One of 30 amazing mathematical topics in daily life*. Gendai Sugakusha.

Source

A longer version of this article appeared in *Plus Magazine* (<http://plus.maths.org>) on 1 March 2006.

CHAPTER 35

Percolating possibilities

COLVA RONEY-DOUGAL AND VINCE VATTER

Imagine you are a farmer with several orchards. Your life is normally pretty stress-free but the one thing that keeps you tossing and turning in bed at night is the threat of blight. An insect or a bird can bring blight into one of your beloved orchards and, once there, the wind can carry it from a tree to its neighbour. So when planting a new orchard you feel conflicted: you want to fit in as many trees as possible, but you also want to plant them far enough apart that blight will not easily spread through your entire orchard.

Let's assume that the trees are planted in straight lines, so they form a square grid as shown on the left in Fig. 35.1. We represent each tree by a dot, as on the right. Depending on how far apart the trees are planted, there will be a probability p of blight spreading from a tree to its neighbour. Lines (called *edges*) represent routes through which the blight could spread from a tree to one of its four neighbours. The closer together the trees are planted, the bigger p will be, but conversely, if they're closer together then we can obviously plant more trees. As a test for which values of p are blight-resistant, we ask: if a tree in the orchard gets blight, what is the chance that a large number of trees get blight?

In Fig. 35.2, we have simulated the possible spread of blight through the orchard by randomly adding each edge with probability p for three different values of p. The picture on the left shows that if $p = 1/4$, then even if a few trees get blight, it will not spread far. In the middle, where $p = 1/2$, the situation is more worrying: there are some large clusters and some small clusters, so one tree getting blight may or may not cause massive ruin. When $p = 3/4$, shown on the right, if even a single tree gets blight, the infection is likely to spread across almost the entire orchard.

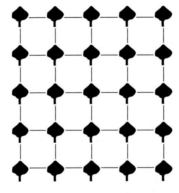 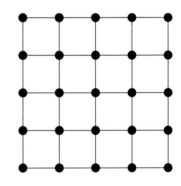

Fig 35.1 An orchard and its mathematical idealisation.

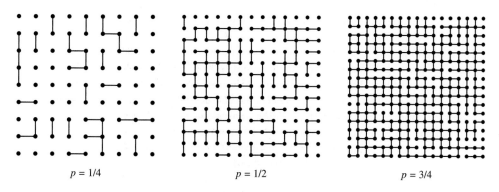

$p = 1/4$　　　　　　　　$p = 1/2$　　　　　　　　$p = 3/4$

Fig 35.2 Simulated blight-spreading plots.

The mathematical study of problems like this is called *percolation theory*. Applications range from the spread of new strains of flu (where we know the probability of an individual passing on the disease to someone they meet, and ask how large an outbreak will be) to the prediction of earthquakes (where an earthquake triggers tiny cracks in the nearby rocks, making it more likely that another earthquake will occur nearby). Physicists often study percolation of extremely large systems, such as water draining through sand. To model these, it is useful to work with an infinite grid. Our orchard question becomes: what is the chance that the infection of a single tree could lead to infinitely many infected trees? If this infinite set could exist, we say that the system *percolates*.

You might think that the chance of percolation would change as the left-hand graph in Fig. 35.3, moving smoothly from 0 to 1 as the probability p that an edge is present increases. But, in fact, infinity plays a cute trick on us. In 1933 the Russian mathematician Andrey Kolmogorov proved his now famous zero–one law. This theorem states that if one has an infinite sequence of events (such as an infinite number of coin flips), then any outcome which cannot be determined by any finite set of the events (such as the coin landing heads infinitely often) occurs with probability 0 or 1. Indeed, in the coin-flipping example, the outcome (infinitely many heads) will occur with probability 1 unless it is a trick coin with two tails, when it will occur with probability 0. In our infinite orchard, the events are the edges being present or absent, and the outcome is the existence of an *infinite* path of infection. The outcome does not depend on any finite set of events, because if there were an infinite path, cutting it into finitely many pieces would still leave at least one infinite part (and possibly more). Therefore, the chance of percolation makes a sudden jump from 0 to 1,

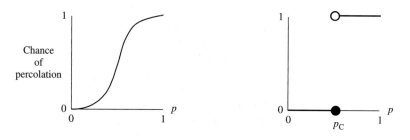

Fig 35.3 Defining the critical probability p_C.

and is never in between, as shown in the right-hand graph in Fig. 35.3. The point where this jump occurs is called the *critical probability*, p_C.

For a long time, physicists were desperate to know the critical probability for percolation on an infinite square grid. Surprisingly, this turns out to be an extremely hard problem. In 1960 Ted Harris showed that the critical probability is at most 1/2. It took another 20 years of effort before a different mathematician, Harry Kesten, was able to establish that the critical probability is precisely 1/2. More recently, Kenneth G. Wilson won a Nobel Prize in Physics for research relating to percolation, and both Wendelin Werner's and Stanislav Smirnov's Fields Medals (these are like Nobel Prizes for mathematics, but only given for work done by mathematicians under 40 years of age) were at least partly for their work in percolation theory.

Harris and Kesten's result shows that if we want a blight-resistant orchard, we should plant our trees far enough apart that the probability of blight spreading from a tree to each of its neighbours is less than 1/2. But would it be possible to pack in more trees by changing the layout of the orchard? In 1964, M. F. Sykes and John Essam showed that the critical probability for a 'honeycomb' arrangement, as in Fig. 35.4, is $1 - 2\sin(\pi/18)$, approximately 0.65. Unfortunately this arrangement means that, for the same distance between the trees, we'll fit fewer in. However, the fact that the critical probability is bigger means that we can plant the trees closer together. There's no single recommendation here for our farmer, as it depends on the type of blight, which determines how much closer the trees can be planted.

Percolation theory has even been used to model the development of galaxies. Ever since the Universe settled down after the Big Bang, every galaxy has contained both stars and regions of gas that might collapse to form stars. However, something needs to happen to trigger this collapse. One likely trigger is an existing star exploding in a supernova, sending a shock wave through space. This can cause a molecular cloud to form, from which new stars can condense. The most massive of those new stars, in turn, will later go supernova themselves, triggering another wave of star formation and causing stars to percolate through space. This effect (taking into account galactic rotation and more) was suggested and modelled by Lawrence Schulman and Philip Seiden in a 1986 article in the journal *Science*. In Fig. 35.5 the left-hand picture is of a real spiral galaxy, whilst the right-hand one was produced by their percolation model of star formation.

If our galaxy were such that it did not reach the critical probability that guarantees percolation of new stars through space and time, the skies above our heads would have slowly darkened over

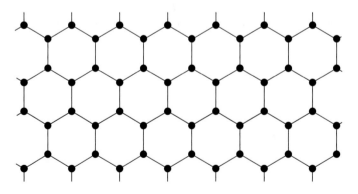

Fig 35.4 Sykes and Essam's honeycomb pattern.

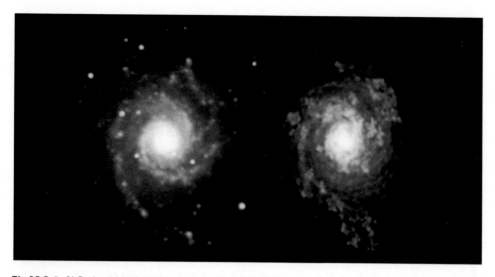

Fig 35.5 (Left) Real and (right) simulated galaxies. Copyright © 1986, American Association for the Advancement of Science.

the millennia, and indeed our own Sun might never have been born. How lucky we are to live in a percolating world!

. . . .

FURTHER READING

[1] Dietrich Stauffer and Ammon Aharon (1994). *Introduction to percolation theory*. Taylor and Francis (2nd edition).
[2] Lawrence Schulman and Philip Seiden (1985). Percolation and galaxies. *Science*, vol. 233, pp. 425–430.

CHAPTER 36

Milestones on a non-Euclidean journey

CAROLINE SERIES

In 1823 the young Hungarian mathematician János Bolyai wrote to his father 'I deduced things so marvellous that I was enchanted. . . . From out of nothing I have created a strange new world.' What was Bolyai's discovery, and what became of his strange new world?

Between about 1820 and 1830 Carl Friedrich Gauss, Nikolai Lobachevsky, and János Bolyai, each working independently, realised that geometry does not necessarily have to obey all the rules laid down by the Greek father of geometry, Euclid. They were interested in what would happen if you didn't assume Euclid's parallel postulate, which says that through a point not on a given line, you can draw exactly one line which never meets the given one. The assumption that there is more than one such line has some strange consequences: the sum of the angles of a triangle is less than 180 degrees (see Fig. 36.1), and any two triangles whose angles are equal are not just similar, but congruent, so their side lengths are also the same. But there were no contradictions, and the three mathematicians became convinced that their new geometry was correct.

One way to think of this new non-Euclidean geometry (often called *hyperbolic geometry*) is that it is geometry on a surface which is saddle-shaped at every point. To see what this means, think of a round crochet mat made by adding successive rings of stitches. Add just the right number of stitches at each stage and the mat stays flat; too few and it tightens into a skull cap; too many and it acquires wavy edges like a leaf of kale. The surface of the kale leaf typifies hyperbolic geometry.

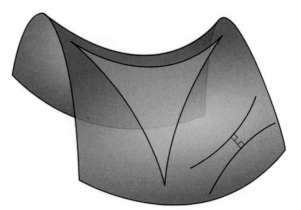

Fig 36.1 The angles in a hyperbolic triangle add up to less than 180 degrees.

By 1860 the young Bernhard Riemann had generalised the idea of hyperbolic geometry to describe curved spaces of any dimension. By the 1880s, despite the best efforts of philosophers to refute these dangerous new ideas, mathematicians were convinced. Geometry did not have to be intrinsically Euclidean, though arguments about its validity and physical significance continued for years.

Hyperbolic geometry soon found unexpected applications. When Einstein developed special relativity in 1905, the mathematical operations he needed were exactly the symmetries of hyperbolic geometry. Later, he realised that Riemann's geometry was just what he needed to work out the theory of general relativity, in which large masses actually curve space.

In the 1920s and 1930s mathematicians found they could use hyperbolic geometry to study the motion of an object sliding on a frictionless surface. As long as the surface wasn't a sphere or a doughnut, hyperbolic geometry told them that although the paths of different objects would be theoretically predictable, to an observer they would look as random as coin tossing. By the 1960s sliding on a hyperbolic surface had become a prototype for what we now call chaos.

Surfaces use geometry in two dimensions. In the 1890s Henri Poincaré wrote down beautiful formulae explaining how hyperbolic geometry works in three dimensions. People soon found abstract crystal structures with the symmetries of 3-dimensional hyperbolic geometry. For 60 years these remained one-off curiosities. Then in the 1970s Bill Thurston began explaining that the curiosities weren't so special at all; they are at the heart of understanding all the possible different types of *3-manifolds*.

A 3-manifold is a 3-dimensional solid, such as a block of metal with complicated holes and tunnels drilled out. Thurston discovered that like a surface, a 3-manifold comes equipped with a natural geometry, which is very often hyperbolic. This suggested a way of understanding and classifying all the possibilities, known as *Thurston's geometrisation conjecture*. His insights created a wonderful new playground for mathematicians. For example, drill a closed knotted loop out of a 3-dimensional sphere. What you have left is a 3-manifold which is naturally hyperbolic, showing that there is an intimate connection between hyperbolic geometry and knots.

One way to try to understand a manifold is to see what closed loops it contains. The Poincaré conjecture, famously solved by Grigory Perelman in 2002, says that if every loop in a 3-manifold can be shrunk to a point without lifting it out of the manifold, then the manifold is a 3-dimensional sphere. Perelman's solution is one of the great mathematical achievements of the early 21st century and led to the proof of Thurston's geometrisation conjecture.

The theory of hyperbolic geometry is still evolving. One of the most fascinating things I have come across recently is that an ingenious approximate version of hyperbolic geometry invented by the brilliant Russian mathematician Mikhail Gromov may give a good way of modelling huge networks like Facebook or the Internet. Preliminary calculations about the routing of messages and transmission speeds show that modelling networks in this way may fit the actual data better than more conventional methods. Being able to work more efficiently with these vast networks would have far-reaching impact in areas like data-mining and brain research. Bolyai would be amazed indeed.

CHAPTER 37

Simpson's rule

SIMON SINGH

I f you are a truly dedicated fan of *The Simpsons* then you might have spotted the name of Springfield's movie theatre, which makes its first appearance in an episode entitled *Colonel Homer* (1992). It is called the 'Springfield Googolplex'.

In order to appreciate this reference it is necessary to go back to 1938, when the American mathematician Edward Kasner was in conversation with his nephew Milton Sirotta. Kasner casually mentioned that it would be useful to have a label to describe the number 10^{100}, or

10,000,000,000,000,000,000,000,000,000,000,
000,000,000,000,000,000,000,000,000,000,000,
000,000,000,000,000,000,000,000,000,000.

The nine-year-old Milton suggested the word *googol*.

In his book *Mathematics and the Imagination*, Kasner recalled how the conversation continued: 'At the same time that he suggested "googol" he gave a name for a still larger number: "googol-plex" '. A googolplex is much larger than a googol, but is still finite, as the inventor of the name was quick to point out. It was suggested that a googolplex should be 1 followed by writing zeros until you get tired.

The uncle rightly felt that the googolplex would then be a somewhat arbitrary and subjective number, so he suggested that the googolplex should be redefined as 10^{googol}. That is 1 followed by a googol zeros, which is far more zeros that you could fit on a piece of paper the size of the observable Universe, even if you used the smallest font imaginable.

These terms – googol and googolplex – have become moderately well known today, even among members of the general public, because the term 'googol' was misspelt and adopted by Larry Page and Sergey Brin as the name of their search engine. Google headquarters is, not surprisingly, called the Googleplex. However, Google was not founded until 1998, so how did the term 'googol' crop up in a 1992 episode of *The Simpsons*?

It is a little-known fact that *The Simpsons* has more mathematically trained writers than any other TV show in history, and these writers have been smuggling dozens of mathematical references into the series over the course of twenty-five years. For instance, the episode *Treehouse of Horror VI* (1995) consists of three short Halloween stories. The third segment, *Homer*[3], demonstrates the level of mathematics that appears in *The Simpsons*. In one sequence alone, there is a tribute to Euler's equation, a joke that only works if you know about Fermat's last theorem, and a reference to the P versus NP problem. All of this is embedded within a narrative that explores the complexities of higher-dimensional geometry.

One of the longest-serving mathematical writers is Al Jean, who worked on the first season and who has since held the senior role of showrunner off and on for many years. Jean was a teenage mathlete, who entered a Michigan mathematics competition in 1977 and who tied for third place out of 20,000 students from across the state. He even attended hot-housing summer camps at Lawrence Technological University and the University of Chicago. These camps had been established during the Cold War in an effort to create mathematical minds that could rival those emerging from the Soviet network of elite mathematics training programmes. As a result of this intense training, Jean was accepted to study mathematics at Harvard when he was only 16 years old.

According to Jean, the mathematical references in *The Simpsons* are introduced after the initial script has been drafted and shared for comments. At this rewrite stage, any mathematician in the room might take the opportunity to sneak in an equation or mathematical concept. Indeed, this is exactly what happened during the writing of *Colonel Homer*: 'Yeah, I was definitely in the room for that. My recollection is that I didn't pitch Googolplex, but I definitely laughed at it. It was based on theaters that are called Octoplexes and Multiplexes. I remember when I was in elementary school, the smart-ass kids were always talking about googols. That was definitely a joke by the rewrite room on that episode.'

Michael Reiss, who has worked alongside Al Jean as a writer ever since the first series of *The Simpsons*, was also a teenage mathematical prodigy. He corresponded with Martin Gardner and earned a place on the state of Connecticut mathematics team. He thinks that the Springfield Googolplex was possibly his gag. When a fellow writer raised a concern that the joke was too obscure, Reiss remembers being very protective: 'Someone made some remark about me giving him a joke that nobody was ever going to get, but it stayed in … It was harmless; how funny can the name of a multiplex theatre be?'

Source

This article is based on an extract from *The Simpsons and their mathematical secrets* by Simon Singh, published by Bloomsbury.

CHAPTER 38

Risking your life

DAVID SPIEGELHALTER

Taking ecstasy is no more dangerous than the addiction to horse riding, or 'equasy'. This assertion was made in 2009 by David Nutt, Chairman to the UK Advisory Council on the Misuse of Drugs, and this and other statements eventually cost him his job. He used statistics to back up his claim, comparing the serious harm suffered by horse-riders as well as ecstasy-users. The furore around his statement had more to do with cultural attitudes to drugs versus horses than statistics, but it raises an interesting question: how do you compare small and lethal risks?

Ideally we need a 'friendly' unit of deadly risk. A suggestion made in the 1970s by Ronald Howard is to use the *micromort*: a one-in-a-million chance of death. This is attractive, since it generally means that we can translate small risks into whole numbers that we can immediately compare. For example, the risk of death from a general anaesthetic (not the accompanying operation) is quoted as 1 in 100,000, meaning that in every 100,000 operations we would expect one death. This corresponds to 10 micromorts per operation.

We can also consider the 18,000 people who died from 'external causes' in England and Wales in 2010. That is, those people out of the total population of 54 million who died from accidents, murders, suicides, and so on. This corresponds to an average of

$$\frac{18{,}000}{54 \times 365} \approx 1$$

micromort per day, so we can think of a micromort as the average 'ration' of lethal risk that people spend each day, and which we do not unduly worry about.

A one-in-a-million chance of death can also be thought of as the consequences of a form of (imaginary) Russian roulette in which 20 coins are thrown in the air: if they all come down heads, then the subject is executed (the chance of this happening is 1 in 2^{20}, which is roughly equal to one in a million).

A measure such as a micromort needs a unit of exposure to accompany it, and we can consider different sources of risk that naturally give rise to different measures of exposure. Of course, we can only quote average risks over a population, which represent neither your personal risks nor those of a random person drawn from the population. Nevertheless, they provide useful ballpark figures from which reasonable odds for specific situations might be assessed.

How you interpret these numbers depends on your philosophical view about what probabilities really are. Do they exist as properties of the external world, in which case the assessed odds can be considered estimates of the 'true' risks, or are they only subjective judgements based on currently available information? We favour the latter interpretation.

Transport

Assuming that the risk remains constant within transport type and over time, we can work out how many micromorts we accumulate for 100 miles travelled on different means of transport in the UK, using 2010 figures:

Walk: 4 micromorts per 100 miles.
Cycle: 4 micromorts per 100 miles.
Motorbike: 16 micromorts per 100 miles.
Car: 0.3 micromorts per 100 miles.

Medical events

What about going into hospital? Each day in England around 135,000 people occupy a hospital bed and inevitably some of these die of their illness. However, not all these deaths are unavoidable: in the year up to June 2009, 3735 deaths due to lapses in safety were reported to the National Patient Safety Agency, and the true number is suspected to be substantially higher. This is about 10 a day, which means an average risk of around 1 in 14,000, assuming few of these avoidable deaths happened to outpatients. So staying in hospital for a day exposes people, on average, to at least 75 micromorts. This is equivalent to the average risk of riding around 500 miles on a motorbike, say between London and Blackpool and back. Of course, these situations tend to concern different sorts of people, but the conclusion is still that hospitals are high-risk places:

Night in hospital (England): 75 micromorts.
Caesarean (England and Wales): 170 micromorts.
Giving birth (average for the world): 2100 micromorts.
Giving birth (UK): 120 micromorts.
General anaesthetic (UK): 10 micromorts.

Comparison of the numbers for transport and health is informative: for example, having a general anaesthetic carries the same risk of death, on average, as travelling 70 miles on a motorbike.

Leisure activities

Here we assume that the risk comes from a specific accident in what is otherwise a safe activity with no chronic ill-effects. It is therefore natural to look at micromorts per activity: we look at micromorts per horse ride, per scuba dive, etc.:

Hang gliding: 8 micromorts per activity.
Running marathon: 7 micromorts per activity.
Scuba diving: 5 micromorts per activity.
Skiing (1 day): 1 micromort per activity.

Living the microlife

All these examples concern sudden deaths, but many risks we take do not kill us straight away: think of all the lifestyle frailties we get warned about, such as smoking, drinking, eating badly, not exercising, and so on. The *microlife* aims to make all these chronic risks comparable by showing how much life we lose on average when we are exposed to them.

A microlife is 30 minutes off your life expectancy – the term comes from seeing that 1,000,000 half-hours is 57 years, around the remaining life expectancy of someone in their 20s.

Here are some things that are associated, on average, with the loss of one microlife of a 30-year-old man:

- Smoking two cigarettes.
- Drinking 7 units of alcohol (e.g. 2 pints of strong beer).
- Each day of being 5 kg overweight.

Of course, this is not a precise consequence of each cigarette smoked or drink drunk: it comes from accumulating the effects of a lifetime of behaviour and averaging it over a whole population. And these are very rough figures, based on many assumptions. If you just smoke two cigarettes and then stop, it is impossible to say what long-term effect it will have on you, for a number of reasons. First, microlives are usually based on comparisons between different people (e.g. people who smoke and people who don't smoke) rather than people who have changed behaviour. Second, we can never know what would have happened had an individual done something different.

There is a simple relationship between change in life expectancy and microlives per day. Consider a person aged around 30 with a life expectancy of 50 years, or 18,000 days. Then a daily behaviour or status that leads them to lose a year of life expectancy (17,500 microlives) means that it's as if they are using up around one microlife every day of their lives.

Live fast, die young

Microlives encourage the metaphor that people go through their lives at different speeds according to their lifestyle. For example, someone who smokes 20 a day is using up around 10 microlives, which could be loosely interpreted as their rushing towards their death at around 29 hours a day instead of 24. This idea of premature ageing has been found to be an effective metaphor in encouraging behaviour change. An example is your 'lung age' – the age of a healthy person who has the same lung function as you; 'heart age' is also becoming a popular concept.

An interesting issue is that life expectancy has been increasing by around three months every year (that is, 12 microlives per day) for the last 25 years. This could be interpreted as follows: although just by living we are using up 48 microlives a day, our death is also moving away from us at 12 microlives a day. Essentially, our health care system and healthier lifestyles are giving us a bonus payback of 12 microlives per day.

Micromorts and microlives

Mathematically, exposing ourselves to a micromort, a one-in-a-million chance of death, corresponds to a reduction of our life expectancy by a millionth. Hence a young adult taking a micromort's acute risk is almost exactly exposing themselves to a microlife (recall that a young man in his 20s is expected to have 1 million half-hours ahead of him). An older person taking the same risk, while still reducing their life expectancy by a millionth, is only perhaps losing 15 minutes' life expectancy. However, acute risks from dangerous activities are not well expressed as changes in life expectancy, and so different units appear appropriate.

Governments also put a value on microlives. The UK National Institute for Health and Clinical Excellence (NICE) has guidelines which suggest the National Health Service pay up to £30,000 if a treatment is expected to prolong life by one healthy year. That's around 17,500 microlives. This means that NICE prices a microlife at around £1.70. The UK Department of Transport prices a 'value of a statistical life' at £1,600,000, which means they are willing to pay £1.60 to avoid a one-in-a-million chance of death, or a micromort. So two government agencies put similar value on a microlife and a micromort.

There is one big difference between micromorts and microlives. If you survive your motorbike ride, then your micromort slate is wiped clean and you start the next day with an empty account. But if you smoke all day and live on pork pies, then your microlives accumulate. It is like a lottery where the tickets you buy each day remain valid forever and so your chances of winning increase every day. Except that, in this case, you really do not want to win.

Source

This article is based on two articles in *Plus Magazine* (<http://plus.maths.org>): 'Small but lethal', which appeared on 12 July 2010, and 'Microlives', which appeared on 11 January 2012.

CHAPTER 39

Networks and illusions

IAN STEWART

Neuroscientists, investigating how the brain processes visual images, have long been intrigued by the way the visual system behaves when confronted with incomplete or contradictory information. Famous examples include the Necker cube, named after the Swiss crystallographer Louis Albert Necker, which alternates between two apparent orientations, the rabbit/duck illusion of Joseph Jastrow, and the cartoonist William Ely Hill's *My wife and my mother-in-law* – see Fig. 39.1. In each case the information received by the brain is consistent with two different interpretations, and the *percept* – the image that the brain perceives – alternates, fairly randomly, between two different choices. We call such perceptions illusions. They are characterised by *incomplete information*.

In similar experiments, *contradictory* information leads to a similar effect, called *rivalry*. Again, the percept alternates between different possibilities, but now some percepts may not be consistent with either visual stimulus. The commonest case is binocular rivalry, where the left and right eyes are exposed to two conflicting images. In a typical modern experiment the psychologist Steven K. Shevell and his co-workers presented each subject's left eye with a vertical grid of pink and grey lines, and the right eye with a horizontal grid of green and grey lines. Four different percepts were reported: the two stimuli, a vertical pink and green grid, and a horizontal pink and green grid. These last two are examples of *colour misbinding*: the perceptual system links the colours together incorrectly. Here, subjects may perceive images that are shown to neither eye (Fig. 39.2).

There are many variations on these themes, and they reveal a variety of strange or unexpected phenomena. The challenge is to explain all of these observations with a coherent theory.

Fig 39.1 Three classic visual illusions. The rabbit/duck illusion image is from the 23 October 1892 issue of *Fliegende Blätter*. The cartoon *My wife and my mother-in-law* appears in *Puck*, vol. 78, no. 2018 (6 November 1915), p. 11.

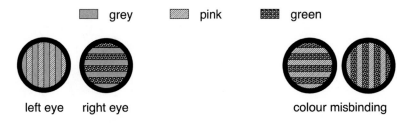

grey   pink   green

left eye   right eye   colour misbinding

Fig 39.2 Shevell's rivalry experiment. (Left) Stimuli presented to left eye and right eye. (Right) Subjects' reported percepts.

One attractive option is to model aspects of perception using networks of neurons. Say that a neuron *fires* if it produces a train of electrical pulses. Two kinds of connection between neurons are important: excitatory connections, in which the firing of a neuron biases connected neurons towards firing as well, and inhibitory ones, in which firing is suppressed.

The simplest network that can model illusions like the Necker cube comprises two identical neurons, one for each percept. In order for a subject to choose one image, each neuron has an inhibitory connection to the other. The crucial variable is the rate at which a neuron fires – the frequency of its pulses. Rate models of this kind can be represented mathematically and they have been widely studied. One model, devised by the mathematician Rodica Curtu, predicts the occurrence of a state in which the rates of firing of the two neurons vary in a periodic manner over time, akin to a wave. But the waves for the two neurons are half a period out of phase – they are in *antisynchrony* (see Fig. 39.3). The standard interpretation is that the observed percept corresponds to whichever neuron has a greater firing rate, so the percepts alternate regularly, as is indeed the case with the Necker cube. In practice the alternation is less regular, probably because of random noise from other neurons in the brain.

Recent work generalises this simple two-cell network to more complex networks that can distinguish between several percepts. These can successfully predict the results of many different experiments on illusions and rivalry, and researchers have even suggested a neural network model for high-level decision making in the brain, based on rivalry.

To see how these ideas work, consider a network modelling Shevell's rivalry experiment (Fig. 39.4). The stimuli have two distinct *attributes*: grid orientation and colour. Each attribute has several possible states, called *levels*: vertical or horizontal for the grid, and pink, grey, or green for the colours. The network represents each attribute as a column of cells (representing levels), coupled to each other by inhibitory connections, which predispose the network to choosing just one level in a 'winner takes all' manner. There are two columns for the colours because each grid

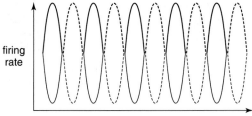

time

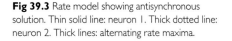

Fig 39.3 Rate model showing antisynchronous solution. Thin solid line: neuron 1. Thick dotted line: neuron 2. Thick lines: alternating rate maxima.

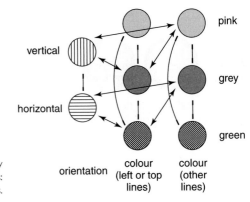

pink

vertical

grey

horizontal

green

orientation

colour
(left or top
lines)

colour
(other
lines)

Fig 39.4 Network modelling Shevell's rivalry experiment. Rectangles: inhibitory connections. Arrows: excitatory connections.

involves two distinct sets of lines of a given colour: lines that start from the left or top, and lines in between these.

In addition, there are excitatory connections, which represent how the stimuli affect the network's responses. For example, the left eye is presented with a vertical grid, grey lines starting from the left, alternating with pink lines. So excitatory connections are added to link these three levels of the corresponding attributes; these predispose the network to respond more strongly to this combination of stimuli. The network can be translated into a system of linked equations – coupled ordinary differential equations to be precise – with one equation per cell.

These *rate equations* involve several parameters, representing how strongly equations are linked together, and other features of the dynamics. The model predicts two distinct states which are periodic over time. In one, the percept alternates between the two stimuli. But in the other, the grid directions alternate but the colours perceived are pink and green – no grey lines. This is exactly what is observed in experiments.

These ideas have now been applied to about a dozen different experiments on rivalry and illusions, using a systematic prescription to construct the network for each experiment. The agreement between theory and experiment is excellent in all cases. Some of the models suggest new predictions, which will enable the ideas to be tested. If they hold up, the network approach will unify a wide range of previously rather puzzling observations and phenomena and offer fresh insights into the mechanisms of visual perception.

FURTHER READING

[1] Richard Gregory (2009). *Seeing through illusions: Making sense of the senses.* Oxford University Press.
[2] Al Seckel (2007). *The ultimate book of optical illusions.* Sterling.
[3] Robert Snowdon, Peter Thompson, and Tom Troscianko (2006). *Basic vision: An introduction to visual perception.* Oxford University Press.

CHAPTER 40

Emmy Noether: Against the odds

DANIELLE STRETCH

When Emmy Noether unexpectedly decided to enter university, German public opinion had just about come round to the fact that some women might, possibly, benefit from higher education. Allowing women to obtain official degrees would have been a bit much, of course, but with the professor's consent they were allowed to sit in on lectures. As Noether's father, Max Noether, was an eminent professor of mathematics at Erlangen, the professors were family friends, and she was able to gain their consent. This was an unexpected direction for any young woman to take at the beginning of the 20th century and was the first of Noether's unlikely successes against the odds in her journey to becoming one of the century's great mathematicians.

Emmy Amalie Noether (Fig. 40.1) was born on 23 March 1882 to a middle-class Jewish family in the small Bavarian town of Erlangen. She was described as a 'clever, friendly, and rather endearing' child who grew up to love parties and dancing as well as absorbing the family atmosphere of mathematics and education. Educational opportunities at the time for a girl growing up in Germany were few, with German schools for middle-class girls being little more than finishing schools.

For three years Noether studied to pass a teacher training programme that would allow her to teach English and French in a girls' school. Then, at the age of 18, she took the unexpected

Fig 40.1 Emmy Amalie Noether (1882–1935).

decision not to become a schoolteacher and instead attend the University of Erlangen, at least in the unofficial capacity the mores of the day allowed.

In 1904, after four years of unofficial study, a relaxation in the rules finally allowed Noether to matriculate at Erlangen officially, and she went on to complete a virtuoso doctorate in 1907. With typical earthy frankness, she later went on to describe her thesis as 'crap'. Women were not allowed to fill academic posts, so Noether spent the next eight years working for the university without pay or position and helping her increasingly frail father with his teaching duties.

David Hilbert, widely considered to be the most influential mathematician of his time, met Noether and her father when they paid an extended visit to the University of Göttingen in 1903. After her doctorate, Hilbert and Felix Klein persuaded Noether to come to Göttingen and then embarked on a long campaign to have her appointed to a faculty position in spite of a Prussian law prohibiting this. Noether was refused a university position but permitted a compromise: she could lecture but only if the lecture was listed under Hilbert's name rather than her own.

Albert Einstein wrote to Klein in 1918, 'On receiving the new work from Fräulein Noether, I again find it a great injustice that she cannot lecture officially.' Finally, in 1919, Noether was granted the lowest faculty rank of *Privatdozent* and, although still unpaid, began teaching under her own name that autumn. A few years later she was appointed to the position of 'unofficial, extraordinary professor': in effect, a volunteer professor without pay or official status, but she was later granted a tiny salary which was barely at subsistence level. When post-war hyperinflation destroyed the value of her small inheritance, Noether had very little left to live on. She pared her life down to the essentials. 'She didn't have very much money, but also she didn't care', her nephew, Herman Noether, explained. During the week, Noether ate the same dinner at the same table in the same cheap restaurant.

Noether made fundamental contributions to both pure and applied mathematics. From a physicist's perspective her most important accomplishment, now known as *Noether's theorem*, described the relationship between symmetries of the laws of nature and conservation principles. For example, in Newtonian mechanics, conservation of momentum arises from the fact that the equations describing ordinary dynamics are the same no matter where you are in space; that there is *space translation symmetry*. Similarly, symmetry in time gives rise to conservation of energy, and rotational symmetry implies the conservation of angular momentum. Noether's theorem also applies to many quantum systems, and this has made her contribution particularly important to the modern development of quantum theory.

In a completely different mathematical direction, Noether published a paper in 1921 called (translated into English) 'Theory of ideals in ring domains', which has had a major influence on the development of modern algebra. Rings are fundamental abstract algebraic structures, and a class of these objects are now called Noetherian rings.

German mathematics, like much else, became highly politicised in the 1930s. Few German academics opposed Hitler's rise to power and one of Noether's research students, Werner Weber, organised a boycott of Professor Edmund Landau's classes because he was Jewish. Hitler then began firing Jewish professors from universities in a bid to remove Germany from the 'satanic power' of the Jews. Noether was one of the first six professors fired from Göttingen because she was both Jewish and politically liberal. The expulsion of Jewish academics devastated the University of Göttingen. Hilbert, asked by a Nazi official about the state of mathematics in Göttingen, replied, 'Mathematics in Göttingen? There really is none any more'.

Noether's friends started a frantic search to find her a university position abroad. Eventually, she was granted a temporary one-year position on a modest salary at a small women's college,

Bryn Mawr, in the United States. Finding her a permanent position proved difficult, as there were too many Jewish refugees and too few places who wanted to hire them. By 1935 enough funds were scraped together to support Noether at a reduced salary for another two years.

Noether then went into hospital to have surgery for the removal of a large ovarian tumour. For a few days it appeared that the surgery had been successful, but then she suddenly lost consciousness and died of what appeared to be a post-operative infection. Sadly, she died in exile in her early fifties at the height of her creativity and powers.

The mathematician and philosopher Hermann Weyl said of Emmy Noether during the troubled summer of 1933 when the Nazis rose to power, 'Her courage, her frankness, her unconcern about her own fate, her conciliatory spirit were, in the midst of all the hatred and meanness, despair and sorrow surrounding us, a moral solace.' The challenges she faced and the climate in which she lived make Noether's mathematical achievements all the more extraordinary.

Source
An earlier version of this article appeared in *Plus Magazine* (<http://plus.maths.org>) on 1 September 2000.

CHAPTER 41

Of catastrophes and creodes

PAUL TAYLOR

The division of mathematics into *pure* and *applied* camps (not to mention the invention of statistics) is a relatively recent phenomenon. This division was firmly entrenched by the middle of the 20th century, when the purist G. H. Hardy could state that the only maths lending itself to practical application was boring, trivial, and hardly 'real mathematics' at all. But even the designation *applied* mathematics suggests a one-way partnership, with scientists or industrialists offering up problems and beneficent mathematicians handing down solutions to the supplicants so that their fields might progress. The reality is much messier and more interesting (see also Chapter 47 for a more philosophical viewpoint).

Mathematicians have gained much from collaborations with other specialists, often being prompted to develop completely novel areas or explore new ideas. The mutually beneficial interaction between mathematicians and others is well illustrated by the development of catastrophe theory. Gaining popularity in the 1970s, it found itself applied to a variety of problems ranging from physics to social sciences and economics, and the applications of catastrophe theory resulted from a collaboration between one of the 20th century's most prominent biologists and an equally prominent mathematician.

To set the context, we start with the biologist. Described in a *Nature* profile as 'the last Renaissance biologist', Conrad Hal Waddington was no stranger to quantitative thinking. He had spent the Second World War engaged in operations research, applying maths and statistics to develop anti-submarine tactics for use in the Battle of the Atlantic, and would later write the definitive treatment of this grossly underappreciated element of the war effort. By 1944 the U-boat threat was dwindling in the face of new technologies and superior Allied air power, and he left to take up the Chair in Genetics at Edinburgh University, where he remained for the rest of his career.

One of Waddington's great interests was in *epigenetics*, a term he had coined himself to describe the interaction of genes with other factors to guide the development of an organism. The cells making up your liver contain the same genes as those which form your skin but express them differently, leading to very different cell types. Waddington imagined the process of determining a cell's fate as a landscape over which a ball could roll freely. Gullies in the landscape represented developmental pathways, referred to by Waddington as *creodes*, with their depth corresponding to their stability against perturbations. A series of branching creodes modelled the cell's transition from pluripotency (a multitude of potential final states) to a specialised final form, and the ball's stopping point represented a cell's fate.

Crucially in this portrayal, the visible effects of mutation on an organism's phenotype are all or nothing. A mutation altering the landscape at the beginning of a trajectory or where a valley splits into two could send the ball to a completely different end point, while one affecting a deep valley in the terrain is unlikely to divert it from its course. Small, cumulative mutations may have no apparent effect, until the last alters the landscape just enough to destabilise a pathway and divert the ball into a neighbouring valley.

Now enter the mathematician. It was this vivid image of trajectories running over a landscape, sometimes merging or vanishing completely as the surface shifted below them, which inspired an eminent pure mathematician, Réné Thom. Thom had won the Fields Medal (the highest award in mathematics and equivalent to a Nobel Prize) for his work in that area of pure mathematics called topology. However, Thom had already been thinking about developmental biology after becoming intrigued by the way in which the topologically complex shapes of young organisms emerged from a single, symmetric egg. The idea of epigenetic landscapes prompted him to think about potential landscapes for dynamical systems and the ways in which equilibria on them shifted as parameters changed.

To take the simplest example, Thom considered the dynamics of the ball described above, which we will assume is at the position x. He thought of the ball as having a potential energy described by a (potential) function $V(x)$ of the form

$$V(x) = x^3 + ax.$$

Here a is a *control parameter*, which you can think of as something that you can vary in an experiment. It is a consequence of Newton's laws of motion that the *force* acting on the particle is minus the derivative of the function $V(x)$. The ball is in equilibrium (i.e. it doesn't change its position) when this force is zero, and this will be at points for which the derivative of $V(x)$ vanishes, i.e. values of x at which $3x^2 + a = 0$. For negative values of a, there are *two* equilibrium values of x. One is stable (the minimum), meaning that the position of the ball does not change very much if you move it a small distance from the equilibrium. The other is an unstable equilibrium (the maximum), meaning that small changes away from equilibrium increase (for an example of an unstable equilibrium, consider the dynamics of a pencil balanced on its end). However, for $a \geq 0$ there are no real solutions to the equation and thus no equilibria. Now imagine that the ball is acted on by gravity and is sitting near to the stable equilibrium point whilst a gradually increases from a negative value. The position of the ball will change very slowly at first, but as a passes through zero the stable and unstable equilibria come together and vanish. The ball will suddenly drop and run off in the negative x-direction at ever-increasing speeds.

The topology leading to this sudden switch in behaviour is known as a *fold catastrophe*. Thom's achievement was to show that for any potential function involving no more than two variables and no more than four active parameters the topological behaviour at discontinuities is described by just seven fundamental types of catastrophe, and to provide a means of studying these singularities.

The next simple catastrophe in Thom's list is the *cusp catastrophe*, shown in Fig. 41.1. Here the two horizontal directions are the control parameters a and b and the height is the dynamic variable corresponding to x. The surface of the cusp catastrophe is the derivative of the quartic potential $V(x) = x^4 + ax^2 + bx$ and satisfies the equation

$$4x^3 + 2ax + b = 0.$$

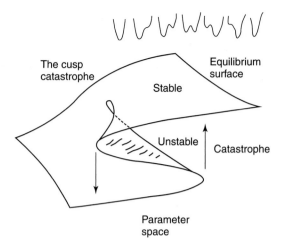

Fig 41.1 Surface of equilibria for a cusp catastrophe (the potential *V* is quartic). The two horizontal directions correspond to the control parameters *a* and *b*, and the vertical is the dynamic variable *x*. The locus of two fold catastrophes forms a cusp when projected onto the parameter plane.

The surface shows the locus of equilibrium values. If the control parameters move from right to left at the back of the surface, all changes in equilibrium positions are small and smooth. On the other hand, if the same change is made at the front of the figure then the changes are smooth until the parameter crosses the bent-back edge of the surface (a fold), where it falls off and drops to the lower sheet: a small change in the parameters has caused a jump in the equilibrium value. This is an example of a *tipping point*, an idea being exploited in the study of climate change today. More complex catastrophes include the elegantly named *swallowtail catastrophe* and the *butterfly catastrophe*, both of which have important applications in optics, structural mechanics, and many other fields.

Critics have disputed the value of catastrophe theory's contributions to science and economics, arguing that it has provided few, if any, results which could not have been proved through more conventional methods. Many today dismiss it as an intellectual fad, and it must be admitted that some of its early proponents were overreaching in claiming its relevance to many problems. Nonetheless, catastrophe theory continues to be used in a variety of fields. Happily, potential-landscape models of epigenetics have enjoyed a renaissance in recent years, using computers to perform calculations which would have been impossible in Waddington's day, so catastrophe theory may yet find application in the problem which inspired it. At present, however, it seems that not only has the development of some beautiful and elegant mathematical results been motivated by collaboration with biologists, but also that mathematics may have profited more from this particular engagement than biology has!

Regardless of catastrophe theory's standing in some applied fields, its internal elegance and success in capturing the public imagination make it one of the most interesting developments in mathematics of the last 50 years. It also stands as a reminder that collaboration with specialists in other fields is not just a service for them, but may also lead to the next big idea in maths.

FURTHER READING

[1] G. H. Hardy (1940). *A mathematician's apology.* Cambridge University Press.
[2] Tim Poston and Ian Stewart (2012). *Catastrophe theory and its applications.* Dover Publications.
[3] Réné Thom (1994). *Structural stability and morphogenesis.* Westview Press.

Source

This article was a runner-up in a competition run in conjunction with *Mathematics Today*, the magazine of the UK's Institute of Mathematics and its Applications.

CHAPTER 42

Conic section hide and seek

RACHEL THOMAS

It's not surprising that today governments and rescue services rely on cutting-edge technology, such as sophisticated electronics and satellite navigation, to keep track of offenders and
undertake search and rescue operations. But it might come as something of a surprise that
ancient mathematics powers these very modern games of hide and seek.

Euclid and Archimedes are just two of the Greek mathematicians that studied *conic sections*:
the shapes created by slicing through a double cone with a flat plane as shown in Fig. 42.1. If the
plane is perpendicular to the axis of the double cone, the shape created is a circle. If it is at an
angle less than the slope of the cone's side, the shape is an ellipse. If the plane is angled parallel to
the side of the cone, the shape is a parabola. If it is angled so as to cut through both cones, we get
a hyperbola.

As well as having a physical description in terms of slicing through a cone, these shapes also
have a very clear geometric description: a circle is the locus of points the same distance from the
centre (the focus of the circle); an ellipse is the locus of points with the same sum of distances
from two foci ($x + y$ equals a constant c in Fig. 42.1); a hyperbola is the locus of points with the
same difference of distances from two foci ($|x - y|$ equals a constant c); and a parabola is the locus
of points equidistant from a focus and a straight line, called the *directrix* ($x = y$).

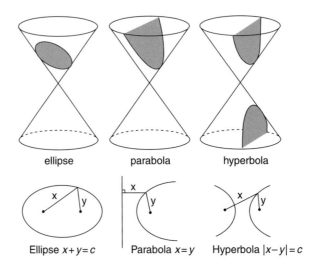

ellipse   parabola   hyperbola

Ellipse $x + y = c$ Parabola $x = y$ Hyperbola $|x - y| = c$

Fig 42.1 The conic sections (the circle can be thought of a special type of ellipse).

Originally the Greeks studied these curves for their fascinating mathematical properties. But over a millennia later it turned out that they had important applications in the real word. In the 16th century Johannes Kepler realised that the planets travelled round the Sun in elliptical orbits. In the early 17th century Galileo constructed one of the first refracting telescopes, using lenses that were the shape of two intersecting hyperbola. Later Newton's designs for reflecting telescopes used a mirrored dish with a parabolic cross section.

Skip forward a few centuries and the mathematical properties of the conic sections now have a very modern application. We are all used to being able to find our location (even if navigationally challenged like myself) thanks to the GPS feature in many mobile phones. GPS satellites, of which there are about nine overhead at any one time, constantly send out signals giving their precise location and the time the signal was sent. The GPS receiver in my phone receives this signal and can calculate my distance, d, from the satellite (equal to the time it takes the signal to reach me multiplied by the speed of light), placing me somewhere on a circle with radius d centred on the location of that satellite. Similarly, the signals from two other GPS satellites place me on two more circles, and the intersection of these three circles pinpoints my location. (Working in two dimensions is a slight simplification but is feasible by assuming we are on the surface of the Earth.)

This method, called *trilateration* (see Fig. 42.2), requires three transmitters (the GPS satellites in this case) to each send a transmission in order to determine someone's location. However, another conic section can make this process more efficient.

Multireceiver radar (Fig. 42.3) uses intersecting ellipses to find a target's location. This time, imagine I transmit a signal to which the target replies. Then, at another location, imagine you have a receiver which picks up the reply. The initial signal I sent has travelled an unknown distance, x, between myself and the target, and the reply has travelled another unknown distance, y, between the target and you. Although we don't know these distances x and y, we do know their sum $x + y$ is the total time from when I sent the signal to when you received the reply, multiplied by the speed of light. Therefore the target is somewhere on an ellipse with you and myself as the foci. By using three receivers we can intersect the three resulting ellipses to find the target's location, this

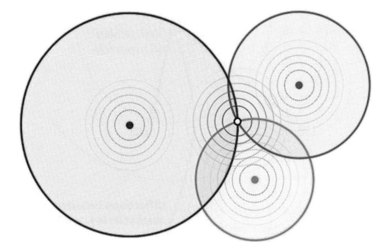

Fig 42.2 The intersection of three circles centred on the GPS satellites pinpoints the target's location.

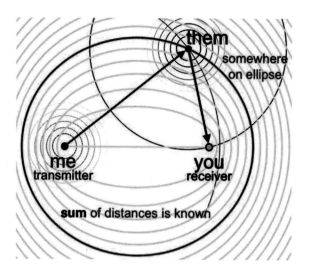

Fig 42.3 Multireceiver radar places the target on an ellipse (given by $x + y = c$, with c being a constant), where the transmitter (me) and the receiver (you) are at the foci.

time using just my lone transmitter and only two transmissions – my initial signal and the target's reply.

One downside of this method is that it assumes the target will immediately reply to any signal we send; and what if we cannot guarantee they would respond? Another conic section provides the answer: the hyperbola.

Hyperbolae allow us to locate a target by silently listening for any signals they transmit. When a transmission is received in two different locations, we know it has travelled an unknown

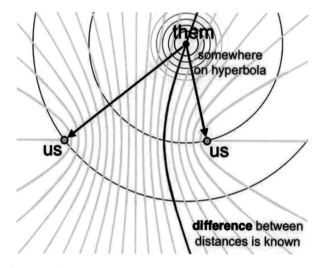

Fig 42.4 Multilateration places the target on a hyperbola (given by $|x - y| = c$, with c being a constant) with the two receivers at the foci.

distance, x, from the target to the first receiver and an unknown distance, y, from the target to the second receiver. The difference between these two distances is then just the difference in flight times of these two paths (that is to say, the measured difference between the times at which the signal was received at the two locations) multiplied by the speed of light. Therefore, the target lies somewhere on a hyperbola with foci at the receivers. The intersection of three such hyperbolae (one for each pair amongst three receivers) locates the target using only the signals the target has transmitted. This covert technique, called *multilateration* (Fig. 42.4), was used in the First World War to locate enemy artillery ranges by listening to the sound of their gunfire. Later, in the Second World War, a hyperbolic navigation system which measured the time delay between two radio signals formed the basis of the Gee navigation system used by the RAF. Nowadays, this 2000-year-old mathematics is still used to find lost souls and uncover hidden enemies.

Source

A version of this article first appeared in *Plus Magazine* (<http://plus.maths.org>). The original article was based on an annual joint London Mathematical Society/Gresham College lecture given by Bernard Silverman, Chief Scientific Adviser to the Home Office, on 20 June 2012.

CHAPTER 43

Sir James Lighthill: A life in waves

AHMER WADEE

Arguably the most eminent UK applied mathematician of his generation, Sir James Light-hill (Fig. 43.1) died as he lived, with perseverance, a unique brand of British eccentricity, and with a childlike spirit of adventure. On 17 July 1998, at the age of 74, he died of a heart attack while attempting to swim around the island of Sark in the English Channel, a feat he had been the first to achieve at the age of 49 and had successfully completed a further five times subsequently. Had he lived, he would have been celebrating his 90th birthday in 2014.

A specialist in fluid mechanics, the practical application of his subject was never far from Lighthill's mind. Every summer he would challenge himself to a different gruelling open-water swim. Part of his reasoning was that as director of the Royal Aircraft Establishment in the early 1960s, he was enthralled by the test pilots who regularly put their life on the line flying experimental vehicles. He felt he needed to show solidarity by putting his own life at the mercy of fluid mechanics.

Few who witnessed his later years can forget this bespectacled, courteous, excitable Colonel Blimp-like figure, or regard him without affection. His idiosyncratic lectures reminded some of the monologues by the British comedian Ronnie Barker, both in the style of delivery and in an uncanny physical resemblance. While listening to Lighthill clearly provided inspiration to his audiences, his use of visual aids could seem from a bygone age. There are many first-hand

Fig 43.1 Sir Michael James Lighthill (1924–1998).

reminiscences telling of overhead projector slides being crammed full of hand-scrawled equations and text, as if he had been given a large box of different-coloured OHP pens for his birthday and needed to show his appreciation by using them all – on the same slide. Nevertheless, the range of topics on which he gave lectures seemed incredible; the fundamentals of space flight, jet noise, the workings of the inner ear, and how fish swim being just a few fields in which he essentially pioneered much of our modern understanding.

Lighthill was born on 23 January 1924 in Paris, France, where his father, 'Bal', was working as a mining engineer. 'Bal' had Alsatian roots and had changed his surname from Lichtenberg in 1917, apparently due to the prevailing anti-German sentiment during the First World War. The family returned to England a few years later and Lighthill was educated at Winchester College, where he was a precocious student; he was a contemporary of another future eminent scientist, Freeman Dyson (see Chapter 17), with whom he remained close friends all his life. Together, they were encouraged to study as much mathematics as they could. By the time they were both 15 years old, they had won scholarships to study at Trinity College, Cambridge. However, they did not go to Cambridge until the age of 17 in 1941, by which time they had already immersed themselves in the material that covered up to Part II of the Mathematical Tripos, the taught mathematics course at Cambridge. During the Second World War, degrees were shortened to be of two years in duration. Lighthill and Dyson, however, only attended lectures for Part III of the Tripos. Needless to say, they sat Part II and graduated with first class honours, and obtained distinctions for Part III in 1943.

Lighthill spent the remainder of the war at the National Physical Laboratory. He was assigned to work with Sydney Goldstein, who was instrumental in convincing Lighthill that fluid dynamics was a rich subject where gifted mathematicians could make a significant contribution. It is well documented that this provided the spark that led to the majority of the major contributions by Lighthill in applied mathematics. After the war Goldstein was appointed to the Beyer Chair of Applied Mathematics at the University of Manchester and persuaded Lighthill to join him as a Senior Lecturer. When Goldstein vacated his position to go to Israel in 1950, Lighthill took up the Beyer Chair at the tender age of 26; he remained at Manchester for 13 years, making fundamental contributions to the theories of supersonic aerodynamics, boundary layers, nonlinear acoustics, vorticity, and water waves.

However, it is possibly in the field of aeroacoustics that Lighthill made his greatest impact, after he published two companion articles entitled 'On sound generated aerodynamically' in the *Proceedings of the Royal Society of London* in 1952 (Part 1) and 1954 (Part 2). Part 1 was truly a pioneering paper, with no references, and provided simple scaling laws for sound intensities in terms of the fluid speed. This had a revolutionary impact on aeronautical engine design, since it showed that the sound intensity scaled with the eighth power of the fluid speed, stressing the importance of reducing the air speed in the jets. Later, it turned out that this scaling law only applied to subsonic jets; the sound intensity from supersonic jets scaled cubically with the fluid speed. The work on aeroacoustics has had far-reaching implications, including in the study of inertia–gravity waves in rotating fluids, and hence has led to a much deeper understanding of atmospheric and ocean dynamics. It has even had an impact in astrophysics in the discovery of the importance of acoustic radiation in stars and cosmic gas clouds. In 1953 Lighthill was elected a Fellow of the Royal Society.

In 1959 Lighthill became Director of the Royal Aircraft Establishment and oversaw several innovative developments. His work on supersonic aerodynamics was instrumental in the use of the technology in passenger aircraft that eventually led to the development of Concorde. Another major development was in vertical take-off and landing (VTOL) technology that led directly to the Harrier 'jump jet'. Although the role was largely administrative, Lighthill continued to publish significant scientific papers, including very early work in the field of biological fluid mechanics; in a seminal paper he outlined all the principal features of what is now known as the standard mathematical model of the dynamics of fish swimming.

Returning to the academic fold in 1964, Lighthill took up the post of Royal Society Research Professor at Imperial College London within the Department of Mathematics. During this time his principal focus was developing theories for application in the fluid mechanics found in wave propagation and in biological systems, the latter being highly influential in subsequent years when he rejoined his alma mater to take up the Lucasian Professorship of Mathematics at the University of Cambridge Department of Applied Mathematics and Theoretical Physics in 1969. Thus he occupied the same chair (academically rather than anatomically speaking) previously held by Isaac Newton, George Stokes, and Paul Dirac. The chair's next incumbent was Stephen Hawking.

After ten years at Cambridge, Lighthill returned to London, this time to University College London (UCL), where he was appointed as Provost. He was a highly popular figure who engaged himself in all aspects of College life. A lifelong devotee of music and an accomplished pianist, Lighthill performed as a soloist for the UCL Chamber Music Society on many occasions. He officially retired in 1989 and spent the subsequent years devoted to full-time research, having relieved himself of considerable administrative burdens.

Throughout his career, Lighthill was a great advocate of communicating scientific findings with clarity, and ceaselessly encouraged the development of younger scientists through his role as either an academic or a journal editor. Indeed, as an editor it was well known that his rates for accepting articles for publication sometimes raised eyebrows from his colleagues. However, closer scrutiny revealed that he spent an inordinate amount of time working with authors to improve articles where he felt there were promising and novel ideas in the work. This is perhaps a salutary lesson for current editors, where the fashion is for scientific journals to increase rejection rates for submitted articles. Although the current policies are designed to increase the quality of the publications, it could be argued that an unwanted side effect is that they greatly encourage risk-averse and perhaps unadventurous research work. Unsurprisingly, Lighthill was showered with international scientific honours, most notably the Timoshenko Medal in 1963, the Royal Medal in 1964, the Elliott Cresson Medal in 1975, the Otto Laporte Award in 1984, and the Copley Medal in 1998; the final one of these was eventually awarded to him posthumously.

In 1959 Lighthill had been the first to propose the idea of setting up a professional institution to uphold the standards and conduct of its members and to provide a voice for the applied mathematics community. He was therefore a powerful driving force behind the founding of the Institute of Mathematics and its Applications (IMA) and served as its first President from 1964 to 1966. He was also a strong advocate that applied mathematics should be treated as a discipline in its own right as distinct from, for instance, pure mathematics or theoretical physics. He truly was a visionary and an extraordinary figure in UK science.

· ·

FURTHER READING

[1] David Crighton (1999). Sir James Lighthill. *Journal of Fluid Mechanics*, vol. 386, pp. 1–3.
[2] Tim Pedley (2001). Sir (Michael) James Lighthill. 23 January 1924–17 July 1998: Elected F.R.S. 1953. *Biographical Memoirs of Fellows of the Royal Society*, vol. 47, pp. 335–356.

CHAPTER 44

Fail safe or fail dangerous?

AHMER WADEE AND ALAN CHAMPNEYS

From the pyramids and skyscrapers to illustrious personalities like Thomas Telford, Isambard Kingdom Brunel, Gustave Eiffel, and Ove Arup – structural engineering has a rich history. Structural engineers are rightly obsessed with safety, and countless measures are taken to make sure their artefacts withstand all kinds of hazard, whether natural or otherwise. For the structural engineer, as the saying goes, failure is not an option. Yet, high-profile catastrophic failures do occur, often blamed on engineers pushing the boundaries of knowledge beyond breaking point or just downright shoddy engineering – examples are the Tay and Tacoma Narrows bridge disasters, several box-girder bridge collapses in the 1960s and 1970s, and the collapse of buildings not directly hit by the planes during the 9/11 attacks in New York.

As in other safety-critical industries, such as aerospace and nuclear power, the civil engineering profession is highly safety-conscious and adept at learning from failure. After each incident, a board of inquiry is set up and the underlying cause of the accident is established. Codes of practice are written and numerous national and international bodies ensure that they are rigorously adhered to. Their aim is to prevent any similar failure from occurring ever again.

Nevertheless, as we shall see, new mathematical insights are beginning to reveal a new way of thinking about structural failure. In our rapidly changing world, with ever-increasing environmental and societal challenges, perhaps we should expect things to fail. Earthquakes, tsunamis, or tropical storms might subject buildings to conditions for which they were just not designed. Moreover, as the world becomes more urbanised, the built environment is likely to be exploited in a manner that is beyond anything that was ever anticipated. The crucial question, then, is perhaps not to predict exactly when a structure would fail: unfortunately, it is inevitable that some will, although just which ones is the subject of huge uncertainties. Rather, we should ask whether a structure is vulnerable to catastrophically dangerous collapse, or whether a failure can be contained. Moreover, if structures are predicted to fail dangerously, is this always a bad thing? We shall see . . .

Structural engineers tend to like the predictable. They like to use elements such as girders, frames, floors, walls, and so on that are designed to undergo deflection; the distance that a structure moves due to an applied force (or *load*) is in strict proportion to the magnitude of that load. This is called *linear* behaviour. For example, if a load of 100 kilonewtons (the weight on Earth of a 10 tonne bus) is applied to a steel girder that results in the deflection of the girder increasing by 1 millimetre in the direction of the load, then, if the girder behaves linearly, doubling the load to 200 kilonewtons would double the deflection to 2 millimetres. However, when loads increase beyond their design levels, this strict proportionality tends to be lost; that is, the structural response becomes *nonlinear*. An example of a nonlinear response would be if our girder deflected by 100 (rather than 10) millimetres when the load was increased to 1000 kilonewtons: the deflection increases disproportionately to the increased load. This generally happens because of the reduction of a structure's *stiffness*, which is the ratio of the load to the deflection.

Mathematics has a lot to say about how nonlinear effects can enhance or worsen stability, going back to *catastrophe theory* in the 1970s (see Chapter 41). In particular, catastrophe theory enables us to predict how a structure is affected by buckling; that is, when it transforms from being in a stable to being in an unstable equilibrium.

A simple example of an unstable equilibrium is a pencil balanced on its point. Theoretically, the pencil should remain upright, but the tiniest disturbance makes it fall over. Once it has fallen over, the pencil is in a stable equilibrium, resting on its side – tiny disturbances then do not significantly affect the position of the object.

Buckling is most likely to occur in structures that are made from slender elements and are compressed, such as the columns and beams that may be found in buildings, bridges, aircraft, offshore platforms, pipelines, or anything that has to withstand significant external load. The level of structural loading at which buckling occurs is known as the *critical* buckling load and is denoted by the symbol P^C. This particular load can in fact be predicted using traditional, tried and tested, linear theory. However, beyond this, to evaluate the so-called *post-buckling* response, inclusion of nonlinear effects is essential and this provides engineers with key information on the existence or otherwise of any residual strength in the structure once buckling has occurred.

Consider Figs 44.1 and 44.2; these represent the response of safe and dangerous mechanical systems, respectively. The spring and link models shown in both figures represent highly simplified (but representative of real structures) mechanical systems that, when loaded by P, reproduce the graphs of load P versus deflection Q; straightforward application of Newton's laws of motion can determine this. In the simplest cases, the geometry of the mechanical systems is 'perfect': initially, when they are not loaded ($P = 0$), they are undeflected ($Q = 0$) and the central springs carry no load. The load is then increased: initially nothing else occurs. However, each mechanical

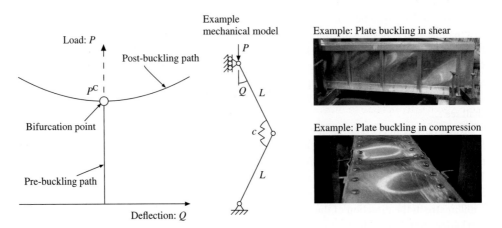

Fig 44.1 Response of a 'safe' mechanical system, where the load P is able to continue increasing beyond the critical buckling load P^C. Left to right: typical load-versus-deflection diagram; a rotational spring and link model that under loading P gives a safe response; real examples in plate girders.

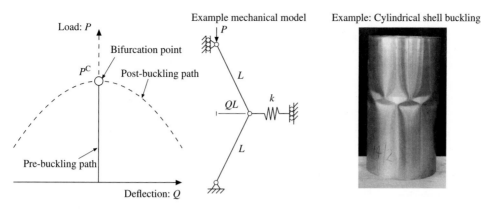

Fig 44.2 Response of a 'dangerous' mechanical system, with P having to reduce after buckling at the critical load P^C. Left to right: typical load-versus-deflection diagram; a longitudinal spring and link model that under loading P gives a dangerous response; a real example, a cylindrical shell.

system encounters a *bifurcation point* when the load P reaches the critical load P^C and the initial undeformed geometry, defined by $Q = 0$, becomes unstable. At this point, the structure effectively has a choice of post-buckling states: in the models shown, with $Q > 0$ the central hinges deflect to the right and with $Q < 0$ the same hinges deflect to the left. In the perfect cases, each system choice has equal probability; only initial system imperfections impose a particular sign for Q.

The nature of the nonlinearities in the system governs whether it is safe or dangerous: a *safe* system can carry loads greater than the critical load, despite buckling occurring (see Fig. 44.1), while a *dangerous* system suffers a catastrophic collapse when it buckles under the critical load P^C, after which it can only bear much smaller loads ($P < P^C$), as shown in Fig. 44.2. In safe systems, the source of the extra strength after buckling is due to geometric effects such as the bending of a buckled surface (for example a thin metal plate) around two axes. Physically, this cannot be achieved without stretching the surface. You can demonstrate this for yourself: try bending a sheet of paper around one axis and you can make a cylinder, fine ... no stretching is required. Try folding the same sheet over the surface of a football, this requires bending around two axes and the sheet cannot cover the surface of the football without gaps appearing, unless of course the sheet is physically stretched out. The stretching of the surface introduces an additional counteracting force that resists the external load P. Hence, as the amount of stretching increases, the load P also has to increase to maintain equilibrium. This type of deformation is seen in the photographs in Fig. 44.1. In dangerous systems, this additional counteracting force is not introduced and the buckled structure has inherently less strength; hence the load has to reduce to maintain equilibrium.

Figure 44.3 shows the effect of imperfections on the behaviour of the systems in Figs 44.1 and 44.2. The *imperfect paths* shown in Fig. 44.3 are curves representing the load-versus-deflection responses of real structures, since it is of course practically impossible to manufacture structural elements (columns, beams, and so on) that are free from defects. In the simple spring and link model examples shown in Figs 44.1 and 44.2, the imperfection, denoted by $Q = \epsilon$ in the corresponding graphs in Fig. 44.3, is an initial misalignment of the central hinge when the springs are unstressed and the structures are unloaded ($P = 0$). Although the imperfection breaks the

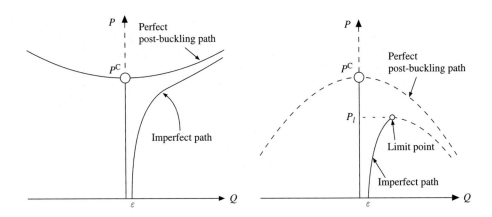

Fig 44.3 Effects of introducing an initial imperfection ϵ into the safe (left) and dangerous (right) mechanical systems.

symmetry of the perfect case, an analysis of the graph for safe structures shows that they can carry loads safely above P^C.

For the dangerous case, however, the equivalent curve reaches a maximum load or *limit point* $(P = P_l)$, where $P_l < P^C$. Moreover, the greater the imperfection size, the greater the difference is between P_l and P^C. Hence, dangerous structures can never actually reach the critical buckling load P^C – immediate failure is likely, and the consequences of buckling are likely to be catastrophic in the usual sense.

Once buckled, structures physically have qualitatively distinctive deformation patterns as demonstrated in the photographs in Figs 44.1 and 44.2. The safe examples buckle into shapes that form some kind of distributed and wavy pattern. Typically, it is a smooth and steady transition – damage may be seen but there is no need to panic! In contrast, in the dangerous case, the structural deformation jumps to a localised rather than distributed form, as seen in the row of dimples shown in the crushed cylinder in Fig. 44.2. This jump tends to release energy suddenly, often heard in the form a crack or bang. For example, listen to the noise when you crush a drinks can by placing it vertically and standing on it. Recent mathematical understanding has shown that the distinction between these two different kinds of post-buckling shapes is general to a wide variety of different buckling problems.

The theory also reveals that more complex structures can show a mixture of these safe and dangerous features. A particularly intriguing phenomenon is known as *cellular buckling*. Imagine a sequence of successive failures like the closing of a concertina. At each buckle, a new cell collapses forming a localised indentation or dimple, and then the geometry of the cell forces the structure to stiffen up again.

This cellular buckling phenomenon can actually be exploited positively, for instance in absorbing the impact energy during a car crash. Essentially, crumple zones embedded in the bodywork of cars are pieces of material that are designed to undergo a sequence of successive buckles. Each buckle results in a localised patch of deformation that soaks up a piece of the energy which would otherwise be transmitted to the occupants of the vehicle. Similar principles apply in plastic egg boxes, which are designed to buckle in such a way as they absorb the impact of being dropped while protecting the delicate eggs inside. The same idea could also be applied in the built environment to protect critical infrastructure such as nuclear power stations or seats of government in the event of a major external hazard such as an earthquake or an explosion.

Structural failure in the form of buckling is often thought of as something to be avoided at all costs. However, this is a simplistic viewpoint. Nonlinear mathematics is beginning to show engineers whether potential failure modes are catastrophic or otherwise. Using this theory, structures can be designed to fail safely or dangerously. Yes, designed to fail *dangerously*! Why on earth might we want that? Well, if the dangerous failure case leads to cellular buckling, then this very process can be exploited to absorb energy and keep other critical components safe.

· ·

FURTHER READING

[1] James Gordon (2003). *Structures: Or why things don't fall down*. DaCapo Press (2nd edition).

[2] Alan Champneys, Giles Hunt, and Michael Thompson (eds) (1999). *Localization and solitary waves in solid mechanics*. Advanced Series in Nonlinear Dynamics, vol. 12. World Scientific.

[3] Giles Hunt, Mark Peletier, Alan Champneys, Patrick Woods, Ahmer Wadee, Chris Budd, and Gabriel Lord (2000). Cellular buckling in long structures. *Nonlinear Dynamics*, vol. 21, pp. 3–29.

CHAPTER 45

Leapfrogging into the future: How child's play is at the heart of weather and climate models

PAUL WILLIAMS

To most people, leapfrog is a traditional children's game. Depending on your age, you probably have fond memories of playing it when you were at school, or perhaps you still play it today. To play, one participant bends over, and another participant runs up from behind and vaults over the first participant's stooped back. Games of this sort have been played since at least the 16th century (see Fig. 45.1).

According to the Oxford English Dictionary, the first recorded use of the word 'leapfrog' dates from 1600, appearing in William Shakespeare's *Henry V*:

If I could win thee at leapfrog,
Or with vawting with my armour on my backe,
Into my saddle,
Without brag be it spoken,
Ide make compare with any.

Leapfrog is played around the world but different countries name it after different leaping animals. It is called 'leapsheep' (*saute-mouton*) in French, 'leaphorse' (*umatobi*) in Japanese, and 'goat' (*capra*) in Romanian.

What has a sixteenth-century children's game got to do with modern mathematics? To mathematical scientists like me, leapfrog is not just a children's game, but also a mathematical technique that is named after the vaulting children. The technique is at the heart of modern weather and climate prediction.

At around the time of the formation of the Institute of Mathematics and its Applications, atmospheric scientists were busy developing the first ever computer models of the weather. These models encapsulate our understanding of the physical laws that govern the atmosphere and ocean, represented as nonlinear mathematical equations. Over time the models have evolved into sophisticated computer codes, containing millions of lines of instructions and requiring hundreds of trillions of calculations a second. Only the world's fastest supercomputers are capable of

Fig 45.1 Detail from *Children's Games* (1560) by the Flemish Renaissance artist Pieter Bruegel the Elder.

performing these calculations. Today, the models are used routinely by national meteorological offices to forecast the weather and to predict the climate.

Children do not have the monopoly on playing leapfrog: weather and climate models do it too. The role of the leapfrog in models is to march the weather forward in time, to allow predictions about the future to be made. In the same way that a child in the playground leapfrogs over another child to get from behind to in front, the models leapfrog over the present to get from the past to the future. In mathematical terms, the time derivative at the present step, which represents the instantaneous rate of change of the atmospheric and oceanic variables, is used to extrapolate forward linearly from the past step to the future step.

The models must proceed in the above manner because, although time in the real world flows continuously from one moment to the next, time in computer simulations is divided up into discrete chunks. In mathematical terms, the differential equations governing the evolution of the atmospheric and oceanic fluid flow are discretised in time, to turn them into algebraic finite-difference equations. The task of forecasting tomorrow's weather is broken up into hundreds of mini-forecasts, each of which advances the prediction by just a few minutes.

Anyone who has ever played leapfrog in the playground will know that it is all too easy to become unstable and fall over. Leapfrogging over a single participant is usually not a problem, but leapfrogging over many participants in quick succession is riskier. The same is true of the mathematical leapfrog in computer models. Each leapfrog step takes the model further into the future but also increases the instability that is inherent in leapfrogging. Eventually, the instability becomes too large and the model is said to crash.

A brilliant Canadian atmospheric scientist called André Robert (1929–1993) fixed this problem 45 years ago, by devising a sort of mathematical glue that holds the model together. The glue is now called the Robert filter and the stickiness stops the instability from growing. How it does this is best described by analogy with the children's game. The filter basically creates subtle forces, which push downwards on the leapfrogger when he or she leaps too high, and which push upwards when he or she starts falling to the ground. The invisible hand of the Robert filter keeps the leapfrogger on a stable trajectory. The filter has been used in most weather and climate models for nearly five decades and it continues to be used widely today.

Unfortunately, although the subtle forces of the Robert filter stabilise the leapfrog technique and prevent the model from crashing, they also introduce errors into the model. These errors are a problem because weather and climate models must be as accurate as possible, given the importance of meteorological events to society and the economy. For example, research published in the *Bulletin of the American Meteorological Society* has calculated that the value of weather forecasts in the USA is $31.5bn annually. The US National Oceanic and Atmospheric Administration has estimated that the cost of underpredicting or overpredicting electricity demand due to poor weather forecasts is several hundred million dollars annually. The US Air Transport Association has calculated that air traffic delays caused by weather cost about $4.2bn annually, of which $1.3bn is potentially avoidable by improved weather forecasts.

In addition to the economic costs of errors in models, there are also societal costs. Many atmospheric and oceanic phenomena represent critical perils for society because of their vast destructive power. Examples include hurricanes, typhoons, mid-latitude wind storms, tsunamis, and El Niño events. These phenomena can lead to various problems, such as flooding, landslides, droughts, heatwaves, and wildfires. Improving the accuracy of weather models is important because it helps to minimise the societal impact of these extreme events. Improving the accuracy of climate models is equally important because their predictions are used in international political negotiations.

For all the above reasons, the errors caused by Robert's 45-year-old invisible hand needed to be fixed. What was needed was a pair of hands, each pushing in opposite directions at slightly different times. Although the force applied by each hand causes an error, the forces applied by the pair of hands are equal and opposite. They cancel each other out, resulting in no net force. This cancellation greatly reduces the errors caused by the Robert filter, and increases the accuracy of the model. In a sense, a different kind of glue is used to prevent the leapfrog instability and hold the model together. Moreover, the method is easy to include in an existing model, requiring the addition of only a couple of lines of computer code.

Since the modified Robert filter was developed, it has been found to increase the accuracy of weather forecasts and ocean simulations significantly. In one atmospheric model, predictions of the weather five days ahead with the modified filter were found to be as accurate as predictions of the weather four days ahead with the original filter. Therefore, the modified filter has added a crucial extra day's worth of predictive skill to tropical medium-range weather forecasts. The modified filter is currently being included in various other atmosphere and ocean models around the world, including models of storms, clouds, climate, and tsunamis. It is somewhat humbling that this mathematical and scientific advance had its roots partly in a simple children's game that I used to play when I was at school. Sometimes, scientific progress really is child's play!

• •

FURTHER READING

[1] Jacob Aron (2010). *Mathematics matters: Predicting climate change*. IMA/EPSRC leaflet, also published in *Mathematics Today*, December, pp. 302–303.

[2] Damian Carrington (2011). How better time travel will improve climate modelling. *The Guardian*, 4 February 2011. <http://www.guardian.co.uk/environment/damian-carrington-blog/2011/feb/04/time-travel-climate-modelling> (site accessed January 2013).

[3] Paul Williams (2009). A proposed modification to the Robert–Asselin time filter. *Monthly Weather Review*, vol. 137, pp. 2538–2546.

Source

This article was the winning entry in a competition run in conjunction with *Mathematics Today*, the magazine of the UK's Institute of Mathematics and its Applications.

CHAPTER 46

Motorway mathematics

EDDIE WILSON

If you own a car, you've probably had this experience: you are driving along the motorway, it's a busy day, but the traffic seems to be flowing fine. All of a sudden, however, you see brake lights ahead of you and everything grinds to a halt. You then inch forward in a queue for maybe 5 or 10 minutes, but when you reach the front of the queue, there is no apparent blockage that would explain the jam – there is no accident, no roadworks, no lane closure. So what caused the jam?

When you reach the front of the queue, you drive off at a good speed but – very often – it is only a few more minutes before you run into the back of another apparently inexplicable queue. For example, if it is summer and you are driving to a popular holiday destination in England such as the West Country or the Lake District, then it is a pattern that might repeat itself all day. Since there is no obvious cause for these jams, they are sometimes called *phantom traffic jams*.

How do we do traffic jam science? These days the road operators try to manage motorway traffic (more on this later), and a by-product of this is a detection system which counts cars and measures their speeds. This system employs electromagnetic induction in loops of wire buried in the road surface every 500 metres or so in each lane. These loops appear as sets of black rectangular outlines on the road surface; don't try to look too closely at these if you happen to be driving, though.

We can use the data from the loop system to build a detailed picture of how traffic jams build up. Figure 46.1 shows a typical schematic, where we have adopted an idea from mathematical physics to look at the motorway in a *space–time* plot. In the plot, the gradient of a vehicle's trajectory has units distance/time and thus represents the speed of the vehicle, which is reduced when it crosses congestion, indicated by grey shading. The congestion pattern typically has two components: a stationary band of congestion, depicted here by the horizontal band, usually located at a busy motorway junction, where the traffic first comes to a halt, and a set of *stop-and-go waves* that come out of the stationary congestion, depicted here by the bands which slope from top left to bottom right. Since this slope is opposite to the direction of vehicle trajectories, these waves travel up the motorway, in the opposite direction to the traffic. So the reason that you don't typically see what caused each jam is that it was created some distance in front of you in the past and it has then travelled back down the motorway towards you like a wave.

In fact, the speed of these waves is almost the same wherever we look around the world: always in the range 15–20 kilometres per hour (km/h). Moreover, individual waves can be remarkably robust; it is not uncommon to have one that lasts for 4–5 hours and thus rolls backwards along the road for up to 100 kilometres.

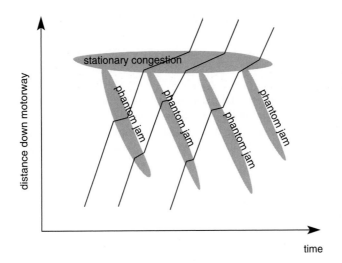

Fig 46.1 Space–time diagram of motorway traffic flow. Vehicles drive forward in space (upwards) and forward in time (to the right) – some typical vehicle paths (called *trajectories*) are plotted as solid lines, thus sloping from bottom left to top right.

To explain phantom traffic jams we need to understand both why the waves move the way they do, which is to say backwards, and what causes the waves in the first place. The second question is much more difficult, so let us deal first with backwards wave propagation.

One key idea is that traffic flow is rather like gas flow in a pipe, where individual vehicles are the equivalent of the gas's molecules. If you were to look down on a motorway from a helicopter you would see all kinds of details of individual driver behaviour, but if the helicopter were to fly high enough you would instead only get an impression of the overall amount of traffic. This is called *density*, measured in cars per length of road. A further gas-like quantity is the average speed of the cars. Then related to the density and speed is the *flow*, which has units of vehicles per unit time, and which is the number of vehicles you would count in, say, one minute, if you were to stand by the side of the road and watch them driving past.

How are speed, flow, and density related? See Fig. 46.2. Firstly, we know that speed tends to go down as the density goes up. This is because drivers become concerned for their safety as the vehicle in front gets too close. A very simple linear model could be

$$v = v_{max}(1 - d/d_{jam}),$$

where v is the speed, v_{max} is the maximum speed, say 70 miles per hour ($\approx$ 112 km/h) in the UK, d is the density, and d_{jam} is the jam density at which everything grinds to a halt, maybe 150 cars per mile ($\approx$ 90 cars per kilometre) per lane of motorway. Furthermore, we know that flow = speed × density. This means

$$flow = v_{max}d(1 - d/d_{jam}),$$

which is an upside-down quadratic; see Fig. 46.2(b) (you can read more about the quadratic equation in Chapter 9). To clarify, if the density is very low, then the speed is high, and it turns out that the product of them is small. Conversely, if the density is very high, then the speed is low, but once again the product of them is small. But at an intermediate density $d = d_{jam}/2$, we obtain the maximum flow rate of $v_{max}d_{max}/4$ – in this case $70 \times 150/4 = 2625$ cars per hour. In fact, this is a little higher than you ever see in practice, because the linear speed–density model is only an approximation of reality.

What's the use of the fundamental diagram, such as in Fig. 46.2? Firstly, the famous British mathematician Sir James Lighthill (see Chapter 43) solved the gas dynamics equations for traffic flow in the 1950s and showed that wave speeds correspond to the gradients of chords that cut the diagram as shown in Fig. 46.2(b). We are skipping a few mathematical steps, but this kind of argument can be used to show why phantom traffic jams go backwards at the speed they do.

Secondly, the fundamental diagram suggests that we might be able to increase the flow of traffic by decreasing its speed (thus increasing its density) – a highly counter-intuitive result. This was the original argument behind *managed motorways*, a concept introduced by the UK Highways Agency and now being rolled out across the UK. Here the speed limit is reduced to 60 and then to 50 miles per hour ($\approx$ 97 and 80 km/h, respectively) in busy conditions.

It is true that reducing the speed limit is a good way of getting rid of jams. However, the mechanism is subtle and to understand why, we need to return to the question of what causes phantom traffic jams in the first place. For this, it turns out that Lighthill's theory is not sufficient and we need to consider *microscopic* simulation models that respect the fine details of each individual driver's behaviour. You can experiment with simple versions of such models at the excellent website maintained by Martin Treiber, which is listed in the further reading section below.

Mathematical analysis of such simulation models has shown that individual driver behaviour tends to be quite *unstable* in some conditions. This means that one only needs to give the flow a small *kick* – corresponding perhaps to a mistake by just a single driver, such as a dangerous lane change. The driver behind the bad lane change hits the brakes, the next driver hits the brakes a little harder, the next driver a little harder, and so on. Soon enough, the kick naturally magnifies to create a phantom traffic jam.

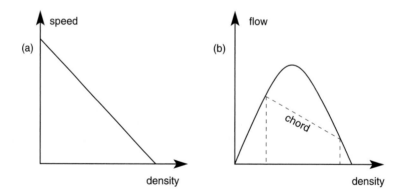

Fig 46.2 Fundamental diagrams of traffic flow: (a) linear speed–density relationship; (b) quadratic flow–density relationship. The speed of the wave that connects traffic at two different densities is given by the gradient of the chord that cuts the flow–density curve. The chord's downward slope implies that the wave travels backwards.

So, to reduce phantom jams highway controllers need to do two things: first, reduce the source of kicks (this is why the second part of managed motorways involves official guidance instructing drivers to avoid making unnecessary lane changes); and second, attempt to stabilise driver behaviour. It is observed that reducing the speed of flow in heavy traffic seems to lead to more stable drivers and fewer phantom jams, but at a fundamental level we still don't understand the details of *why* this is. It remains the subject of ongoing research. Nevertheless, next time you are stuck in a jam on the motorway, you may care to observe the precise pattern of stop–go waves and, instead of looking for a scapegoat, ponder a little on the nature of traffic and its inherent instability.

• •

FURTHER READING

[1] James Lighthill and Gerald Whitham (1955). On kinematic waves II: A theory of traffic on long crowded roads. *Proceedings of the Royal Society A*, vol. 229, pp. 317–345.

[2] Highways Agency. *Managed motorways.* <http://www.highways.gov.uk/our-road-network/managing-our-roads/improving-our-network/managed-motorways/> (site accessed 11 May 2013).

[3] Martin Treiber and Arne Kesting (2011). *Microsimulation of road traffic flow.* <http://www.traffic-simulation.de/> (site accessed 11 May 2013).

CHAPTER 47

The philosophy of applied mathematics

PHIL WILSON

I told a guest at a recent party that I use mathematics to try to understand migraines. She thought that I ask migraine sufferers to do mental arithmetic to alleviate their symptoms. Of course, what I really do is use mathematics to understand the biological causes of migraines. Such work is possible because of a stunning fact we often overlook: the world can be understood mathematically. The partygoer's misconception reminds us that this fact is not obvious. In this article I want to discuss a big question: 'why can maths be used to describe the world?' Or, to extend it more provocatively, 'why is applied maths even possible?' To do so we need to review the long history of the general philosophy of mathematics – what we will loosely call *metamaths*.

Before we go any further we should be clear on what we mean by applied mathematics. We will borrow a definition given by an important applied mathematician of the 20th and 21st centuries, Tim Pedley, the G. I. Taylor Professor of Fluid Mechanics at the University of Cambridge. In his Presidential Address to the Institute of Mathematics and its Applications in 2004, he said 'Applying mathematics means using a mathematical technique to derive an answer to a question posed from outside mathematics.' This definition is deliberately broad – including everything from counting change to climate change – and the very possibility of such a broad definition is part of the mystery we are discussing.

The question of why mathematics is so applicable is arguably more important than any other question you might ask about the nature of mathematics. Firstly, because applied mathematics is *mathematics*, it raises all the same issues as those traditionally arising in metamaths. Secondly, being *applied*, it raises some of the issues addressed in the philosophy of science. However, let us now turn to the history of metamaths: what has been said about mathematics, its nature, and its applicability?

Metamathematics

What is it that gives mathematical statements truth? Metamathematicians interested in such foundational questions are commonly grouped into four camps.

Formalists, such as David Hilbert, view mathematics as being founded on a combination of set theory and logic and to some extent view the process of doing mathematics as an essentially meaningless shuffling of symbols according to certain prescribed rules.

Logicists see mathematics as being an extension of logic. The arch-logicists Bertrand Russell and Alfred North Whitehead famously took hundreds of pages to prove (logically) that one plus one equals two.

Intuitionists are exemplified by L. E. J. Brouwer, a man about whom it has been said that 'he wouldn't believe that it was raining or not until he looked out of the window' (according to Donald Knuth). This quote satirises one of the central intuitionist ideas, the rejection of the *law of the excluded middle*. This commonly accepted law says that a statement (such as 'it is raining') is either true or false, even if we don't yet know which one it is. By contrast, intuitionists believe that unless you have either conclusively proved the statement or constructed a counter-example, it has no objective truth value.

Moreover, intuitionists put a strict limit on the notions of infinity they accept. They believe that mathematics is entirely a product of the human mind, which they postulate to be only capable of grasping infinity as an extension of an algorithmic one–two–three kind of process. As a result, they only admit *enumerable operations* into their proofs, that is, operations that can be described using the natural numbers.

Finally, *Platonists*, members of the oldest of the four camps, believe in an external reality or existence of numbers and the other objects of mathematics. For a Platonist such as Kurt Gödel, mathematics exists without the human mind, possibly without the physical Universe, but there is a mysterious link between the mental world of humans and the Platonic realm of mathematics.

It is disputed which of these four alternatives – if any – serves as the foundation of mathematics. It might seem like such rarefied discussions have nothing to do with the question of applicability, but it has been argued that this uncertainty over foundations has influenced the very practice of applying mathematics. In *The loss of certainty*, Morris Kline wrote in 1980 that 'The crises and conflicts over what sound mathematics is have also discouraged the application of mathematical methodology to many areas of our culture such as philosophy, political science, ethics, and aesthetics [. . .] The Age of Reason is gone.' Thankfully, mathematics is now beginning to be applied to these areas, but we have learned an important historical lesson: there is to the choice of applications of mathematics a sociological dimension sensitive to metamathematical problems.

What does applicability say about the foundations of maths?

The logical next step for the metamathematician would be to ask what each of the four foundational views has to say about our big question. Let us take a different path here by reversing the 'logical' next step: we will ask 'what does the applicability of mathematics have to say about the foundations of mathematics?'

So what can a formalist say to explain the applicability of mathematics? If mathematics really is nothing other than the shuffling of mathematical symbols then why should it describe the world? What privileges the game of maths to describe the world rather than any other game? Remember, the formalist must answer from within the formalist world view, so no Plato-like appeals to a deeper meaning of maths or hidden connection to the physical world are allowed. For similar reasons, the logicists are left floundering, for if they say 'well, perhaps the Universe is an embodiment of logic', then they are tacitly assuming the existence of a platonic realm of logic which can be embodied. Thus for both formalists and non-platonist logicists the very existence of applicable mathematics poses a problem apparently fatal to their position.

By contrast, the central idea of our third proposed foundation, intuitionism, concerning the enumerable nature of processes in the Universe, appears to be deduced from reality. The physical world, at least as we humans perceive it, seems to consist of countable things, and any infinity we might encounter is a result of extending a counting process. In this way, perhaps intuitionism is derived from reality, from the apparently at-most-countably-infinite physical world. It appears that intuitionism offers a neat answer to the question of the applicability of mathematics: it is applicable because it is derived from the world.

However, this answer may fall apart on closer inspection. For a start, there is much in modern mathematical physics, including for example quantum theory, which requires notions of infinity beyond the enumerable. These aspects may therefore lie forever beyond the explicatory power of intuitionistic mathematics.

But more fundamentally, intuitionism has no answer to the question of why non-intuitionistic mathematics is applicable. It may well be that a non-intuitionistic mathematical theorem is only applicable to the natural world when an intuitionistic proof of the same theorem also exists, but this has not been established. Moreover, although intuitionistic maths may seem as if it is derived from the real world, it is not clear that the objects of the human mind need faithfully represent the objects of the physical universe. Mental representations have been selected for over evolutionary time, not for their fidelity, but for the advantage they gave our forebears in their struggles to survive and to mate.

Formalism and logicism have failed to answer our big question and the jury is out on whether intuitionism might do so. What, then, of platonism?

Platonists believe that the physical world is an imperfect shadow of a realm of mathematical objects (and possibly of notions like truth and beauty as well). The physical world emerges, somehow, from this platonic realm, is rooted in it, and therefore objects and relationships between objects in the world shadow those in the platonic realm. The fact that the world is described by mathematics then ceases to be a mystery, as it has become an axiom.

But even greater problems then arise: why should the physical realm emerge from and be rooted in the platonic realm? Why should the mental realm emerge from the physical? Why should the mental realm have any direct connection with the platonic? And in what way do any of these questions differ from those surrounding ancient myths of the emergence of the world from the slain bodies of gods or titans, the Buddha-nature of all natural objects, or the Abrahamic notion that we are 'created in the image of God'?

The belief that we live in a divine Universe and partake in a study of the divine mind by studying mathematics and science has arguably been the longest-running motivation for rational thought, from Pythagoras, through Newton, to many scientists today. 'God', in this sense, seems to be neither an object in the world nor the sum total of objects in that physical world, nor yet an element in the platonic world. Rather, 'god' is something closer to the entirety of the platonic realm. In this way, many of the difficulties outlined above which a platonist faces are identical with those faced by theologians of the Judaeo-Christian world – and possibly of other religious or quasi-religious systems.

The secular icon Galileo believed that the 'book of the Universe' was written in the 'language' of mathematics – a platonic statement begging an answer (if not the question) if ever there was one. Even non-religious mathematical scientists today regularly report feelings of awe and wonder at their explorations of what feels like a platonic realm – they don't invent their mathematics, they discover it.

The hypothesis that the mathematical structure and physical nature of the Universe and our mental access to study both are somehow a part of the mind, being, and body of a 'god' is a considerably tidier answer to the questions of the foundation of mathematics and its applicability than those described above. Such a hypothesis, though rarely called such, has been found in a wide variety of religious, cultural, and scientific systems of the past several millennia. It is not natural, however, for a philosopher or scientist to wholeheartedly embrace such a view (even if they may wish to), since it tends to encourage the preservation of mystery rather than the drawing back of the obscuring veil.

We seem to have reached the rather depressing impasse in which none of the four proposed foundations of mathematics can cope unambiguously with the question of the applicability of mathematics. But I want you to finish this essay instead with the idea that this is incredibly good news! The teasing out of the nuances of the big question – why does applied mathematics exist? – is a future project which could yet yield deep insight into the nature of mathematics, the physical Universe, and our place within both systems as embodied, meaning-making, pattern-finding systems.

Source

An earlier version of this article appeared in *Plus Magazine* (<http://plus.maths.org>) on 24 June 2011.

CHAPTER 48

Mighty morphogenesis

THOMAS WOOLLEY

O n 7 June 1954 the world lost one of its greatest minds. Persecuted for his homosexual life-style and betrayed by the very country he had helped to protect, Alan Mathison Turing, OBE, FRS (Fellow of the Royal Society), reportedly took his own life by eating an apple that had been laced with cyanide. By the time he died at the age of 41, he had revolutionised the fields of logic, computation, mathematics, and cryptoanalysis. We can only imagine what more he could have offered if he had lived for another 20 to 30 years.

Although a sad story, this article is not intended to be a mawkish recounting of the life of Turing or a damning critique of the society and government that condemned him. Instead, I intend to present a celebration of his work and focus on one of his least-known theories about the construction of biological complexity. This work was so far ahead of its time that it took another 30 to 40 years before it was fully appreciated and, even today, it still provides new avenues of research.

Turing's idea

Turing was interested in how pattern and structure could form in nature, particularly because such patterns seem to appear from uniform surroundings. For example, if one looks at a cross-section of a tree trunk, it has circular symmetry. However, this is broken when a branch grows outwards. How does this spontaneous symmetry breaking occur? Turing proposed that this could arise if a growth hormone could be in some way distributed asymmetrically, leading to more hormone concentrated on one part of the circle causing extra growth there. In order to achieve this asymmetry, he came up with a truly ingenious and counter-intuitive theory.

In order to produce patterns, he coupled two components together that individually would lead to no patterning. Firstly, he considered a stable system of two chemicals. The chemicals reacted in such a way that if they were put in a small container the system eventually produced a uniform density of products, and thus no patterns. Secondly, he added the mechanism of diffusion. This meant that the chemicals could move. Incredibly, he showed that if the chemicals were put in a larger container, diffusion could then make the equilibrium state unstable, leading to a spatial pattern. This process is now known as a *diffusion-driven instability*.

To see how remarkable this is, consider putting a spot of ink in water, without stirring. The ink will diffuse throughout the water and eventually the solution will be one shade of colour. There will be no patches that are darker or lighter than the rest. However, in Turing's mechanism diffusion does create patterns. When this happens, the system is said to have *self-organised* and the resultant pattern is an *emergent property*. In this respect, Turing was many years ahead of his time: he showed that understanding the integration of the parts of a system played a role as important as the identification of the parts themselves (if not more so).

Turing termed the chemicals in his framework *morphogens* and hypothesised that cells would adopt a certain fate if the individual morphogen concentrations breached a certain threshold value. In this way, the spatial pattern in morphogen concentration would serve as a pre-pattern to which cells would respond by differentiating accordingly. Typical 1- and 2-dimensional patterns are shown in Fig. 48.1.

We provide an illustration for this counter-intuitive notion based on one of Turing's own (slightly imperialistic) analogies, involving cannibals and missionaries living on an island. Cannibals can reproduce and, without missionaries around, their population increases naturally. Missionaries are celibate, but they can increase their number by converting cannibals. On a small enough island, stable populations of cannibals and missionaries can occur.

However, suppose that the island is made larger and the missionaries use bicycles to get around, making them 'diffuse' faster than the cannibals. This has the potential to destabilise the equilibrium. Previously, if a small excess of cannibals occurred in one place, this excess would grow

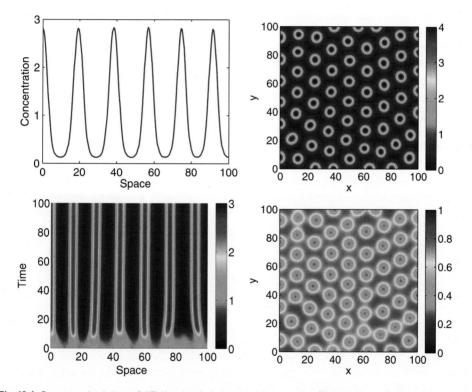

Fig 48.1 Computer simulations of diffusing chemicals that are able to produce Turing patterns. (Left graphs) One-dimensional simulations (aka stripes). The top image shows the final steady state of one of the chemical species. The bottom image shows the temporal evolution. (Right graphs) Two-dimensional simulation (aka spots). The top image shows the final steady state of one of the chemical species and the bottom image shows the final steady state of the other species.

as cannibals reproduced. Now, however, missionaries can quickly spread around the cannibal region. Through conversion, they stop the cannibals from spreading and also produce new missionaries. These can then cycle off to other regions to surround other pockets of cannibalism. If all the parameters are just right, a spotty pattern can emerge: if you painted missionaries green and cannibals red, you'd see red spots surrounded by green spots.

Real-life human interaction is not as simple, but such analogies emphasise that Turing's ideas are not restricted to developmental biology and can be found operating in spatial ecology as well as other diverse areas of nature.

Was Turing right?

One key prediction that comes directly from the theory is important when considering tapered domains, for example the tail of an animal. The theory explains how the patterning on the animal's body arises, but it also predicts that one should see a simplification of the pattern when going from a large domain to a small domain. For instance, a spot pattern on a large domain can transform to a stripe pattern, which is indeed observed in many cases (see Fig. 48.2). This means that animals with striped bodies and spotted tails have coat patterns that are not an emergent property of a simple Turing model. Moreover, if an animal is such that it has a plain body colouring, then its tail should also be plain according to Turing's mechanism. In Fig. 48.2, we see that the cheetah is an excellent example of Turing's model, but that the lemur has, unfortunately, no respect for this mathematical theory. In the latter case, a potential explanation could be that the parameters governing the Turing mechanism are different in the tail from the ones in the body.

Turing proposed that cells responded to morphogens, and patterns emerged as they differentiated in response to differing concentrations of these morphogens. While morphogens have since been discovered, it is still a matter of strenuous debate as to whether morphogens form patterns as Turing envisioned. There is tentative evidence that biological examples of Turing morphogens

Fig 48.2 Pigmentation patterns observed on (a) the cheetah (image: Felinest), and (b) the ring-tailed lemur (with the author).

Fig 48.3 Illustrative examples of chemical Turing patterns for the CIMA (chloride–iodide–malonic acid) reaction within a continuously fed open gel reactor. The gel is loaded with starch, which changes from yellow to blue if sufficient iodide concentrations establish during the CIMA chemical reaction.

exist; however, the jury is still out. In chemistry, it has been shown that Turing patterns can arise, as in the now famous CIMA (chloride–iodide–malonic acid) reaction (Fig. 48.3).

The impact of Turing

During his short life, Turing changed the way people thought in many different fields. Here, we glimpsed only one of his many great ideas: that of producing biological complexity. Turing's work has inspired modelling efforts for describing self-organisation built on different biological hypotheses, leading to the idea of patterning principles and developmental constraints.

However, the model has raised more than its fair share of controversy. Indeed, it is ironic to note that the model, which was developed during an era in which biology followed a very traditional classification route of list-making activities, risks being lost in the present era of data generation and collection. Although all models, by definition, are wrong as they represent a simplification of reality, in biology, the worth of a model must be measured in the number of experiments it motivates that may not have been done otherwise, and in the amount it changes the way experimentalists think. By this measure, Turing's 1952 paper is one of the most influential theoretical papers ever written in developmental biology.

Source
Selected as an outstanding entry for the competition run by *Plus Magazine* (<http://plus.maths.org>).

CHAPTER 49

Called to the barcode

ANDREW WRIGLEY

Barcodes are an integral part of our daily consumer life. They allow machines to read data using an optical scanner, although they have now evolved so that printers and smartphones have the software required to interpret them. Barcodes were first used to label railway carriages, but they became commercially successful when they were used to automate supermarket checkout systems, and now they are ubiquitous.

I have to admit to having my own name called to the very first scanning of a product with a Universal Product Code (UPC) barcode, which took place at a supermarket in Ohio, USA, on 26 June 1974. It was a 10-pack of Wrigley's Juicy Fruit chewing gum costing 67 cents and is famous for no other reason than it was the first item picked out of the trolley of shopper Clyde Dawson. Unfortunately, he never got to enjoy the product, as it was then placed on display at the Smithsonian Institute's National Museum of American History.

The EAN-13 barcode (originally European Article Number) is a 13-digit barcoding standard which is a superset of the original 12-digit UPC system developed in the United States. The barcode consists of a scannable strip of black bars and white spaces, above a sequence of 13 numerical digits, just like the pattern behind the zebra in Fig. 49.1.

With ten possibilities for each of the 13 digits there will be

$$\underbrace{10 \times 10 \times 10 \times 10 \times \ldots \times 10}_{13 \text{ times}} = 10^{13},$$

or ten trillion possible unique barcode numbers using this system. The scanning machines read the black and white lines in their proportion rather than individual size so they can be scaled up or down.

In 1978 *Mad Magazine* once carried a giant barcode across its front cover with the message 'Hope this issue jams every computer in the world'; it didn't. This was partly due to the way that barcode readers work. In the barcode strip, each of the numerical digits is encoded in binary form with a black line representing the symbol 1 and a white space representing 0. Cleverly, the first numerical digit determines a particular way that the next six digits are converted into binary. This is different from the way the last six digits are converted. There is also a characteristic pattern of two vertical lines at the start, the middle, and the end of the strip. This way, an automatic barcode reader can work out the places to start and end reading and in which direction to read the code for itself. Thus barcodes are designed to be read only one way round, which would otherwise cause confusion unless the 13 digits were palindromic.

What each number in a barcode represents is also interesting. The first three digits, known as the prefix, indicate the country of manufacture; anything in the range 930–939 represents

Fig 49.1 A rather confused zebra. © Tim Flach/Getty Images.

Australia, for example, and the special codes 978 and 979 represent *bookland*, which is inhabited by the 13-digit ISBN numbers that uniquely classify all published books. The next few digits indicate the company name, the next few are the product, and the last digit is a check, which picks up any errors in scanning or from manual entry.

The check digit is calculated as follows:

1 The digits in the even numbered positions (2nd, 4th, 6th, 8th, 10th, and 12th) are added and the result is multiplied by 3.

2 The digits in the odd numbered positions (1st, 3rd, 5th, 7th, 9th, and 11th) are then added to this number.

3 The remainder is found when this number is divided by 10 (that remainder is called the number modulo 10).

4 The check digit is 10 minus this remainder.

To illustrate this, consider the barcode on a jar of Vegemite, a popular sandwich spread, similar to the British Marmite. It has a 13-digit barcode number of

9 300650 658516.

First, 9 30 represents Australia as mentioned earlier, where Vegemite is made. Then 0650 65851 gives unique information about the manufacturer, Kraft Foods, and this particular product. Finally, 6 is the check digit. Let us check that it works, by carrying out the above steps.

1 Summing the *even* positions and multiplying by 3, we get

$$(3 + 0 + 5 + 6 + 8 + 1) \times 3 = 69.$$

2 Adding the *odd* positions to this gives

$$69 + (9 + 0 + 6 + 0 + 5 + 5) = 94.$$

3 Dividing this number by 10 gives

$$94/10 = 9, \text{ remainder 4.}$$

4 The check digit should therefore be $10 - 4 = 6$, which it is.

While EAN is the most universal, there are over three hundred versions of the barcode in use today and this is set to grow.

The QR code (quick response code) was first used in Japan in 1994 to track car components in factories. It uses a 2-dimensional array of black or white pixels to store much more information than the vertical lines of a conventional barcode. The characteristic squares in the bottom left, top left, and top right corners are locator patterns, which serve the same function as the vertical lines denoting the ends and middle of an EAN barcode. Every other black or white pixel stores information in binary form. Rather than just a single check digit, it contains three separate embedded mathematical tricks for detecting reading errors. Readers of the QR code are now present in smartphones and have led to new marketing strategies, as people scan the black and white squares and are then taken to a website and consequently recorded in a database.

Fig 49.2 A coin from the Royal Dutch Mint.

Fig 49.3 The Wrigley barcode.

In 2011, the Royal Dutch Mint issued the world's first official coin with a QR code to celebrate the centenary of its current building in Utrecht; see Fig. 49.2. The coin could be scanned by a smartphone, which then linked the user to the Royal Mint website.

In 2012, the science fiction writer Elizabeth Moon proposed that everyone should be barcoded at birth with an implanted chip. This caused something of an outcry at the time, but in the future that means of identification may well replace fingerprints, facial recognition, and retinal scans (for an example, see Fig. 49.3). Of course, we already chip our pets, and the smart cards that we rely upon every day are already a step in that direction.

. . . .

FURTHER READING

[1] Roger Palmer (1999). *The bar code book*. Helmers Publishing (3rd edition).
[2] makebarcode.com. Barcode basics. <http://www.makebarcode.com/info/info.html>.

Source
Selected as an outstanding entry for the competition run by *Plus Magazine* (<http://plus.maths.org>).

CHAPTER 50

Roughly fifty-fifty?

GÜNTER M. ZIEGLER

ave you ever noticed that there are often two consecutive numbers in a lottery draw? For example, the first 2013 draw of the British National Lottery, in which 6 numbers are chosen from 49, produced the numbers 7, 9, 20, 21, 30, 37, with the strange occurrence of the *twins* 20 and 21. On the other hand, the last draw of 2012 didn't have such twins; it produced 6, 8, 31, 37, 40, 48. 'No,' you might say, 'I didn't notice that, because I don't play the lottery.' In my role as a professional mathematician, I could say 'Fine, how wise of you; after all, the lottery is just a tax on people who are bad at maths.' But that still doesn't answer the question 'How often do consecutive pairs occur?' If you look at the results of previous draws of the National Lottery at <http://www.national-lottery.co.uk> (or at similar statistics for German, or Canadian, or other lotteries), you will discover quickly that there's a twin in roughly half of the draws. So it's fifty-fifty? Well, I could say, let's bet one thousand pounds that there will be no twins next Saturday! No, you could say, whoever wants to bet is trying to cheat. And, indeed, if you know your maths (which in this case means knowing your binomial coefficients; the standard pocket calculator doesn't tell you), then you can figure out that the probability is indeed roughly fifty-fifty, but not exactly. Indeed, the probability of twin numbers appearing in a draw is roughly 49.5%. So if I bet on 'no twins', the odds are slightly in my favour, although probably most people think that the chance of a twin is a lot less than fifty-fifty.

Let's do the maths! Of course, I assume (is this also fifty-fifty odds, roughly?) that you know and like your binomial coefficients. Of course, every child should learn them, even if it's not as dramatic as this little scene from Walter Moers' *13.5 lives of Captain Bluebear*:

'Ah,' I said.
'You see!' it cried. 'Anyone who can say "Ah" can learn to say "binomial coefficient" in no time at all!'

This is how the Babbling Billows teaches little Captain Bluebear to speak. Learning to speak, learning to fly, learning to calculate, learning to deal with binomial coefficients – is it easy? It is child's play! If you have forgotten, you might like a reminder: the binomial coefficients for non-negative integers n and k are defined as

$$\binom{n}{k} = \frac{n!}{(n-k)!k!},$$

where $r! = 1 \times 2 \times 3 \times \cdots \times (r-1) \times r$ is the factorial and $0! = 1$. The binomial coefficient describes the number of ways in which k objects can be chosen from n objects, when the order in which they are chosen does not matter. For this reason, the binomial coefficients are sometimes denoted $^{n}C_{k}$, and pronounced 'n choose k'.

Returning to our lottery problem, how many different ways are there to draw six numbers from 49? Well, that's the binomial coefficient $\binom{49}{6}$, which can be calculated exactly using the formula above to be

$$\binom{49}{6} = 13,983,816.$$

But how many ways are there to draw six numbers from 49 without any two consecutive numbers? This may be seem like a hard problem at first, but here's a wonderful trick. In order to get six numbers without consecutives, we first draw six numbers out of 44 numbers, namely out of the numbers $1, 2, \ldots, 43, 44$, and for this there are

$$\binom{44}{6} = 7,059,052$$

different choices. Once we have these six numbers, say $1, 4, 5, 23, 29, 42$ (sorted), we add 0 to the first, but we add an 'offset' of 1 to the second, 2 to the third, and so on, so the last number gets an offset of 5 added. This produces six numbers in the range from 1 to 49, and these numbers won't have consecutives, because in our sequence of six numbers the next number is larger and it also gets a larger offset, so the gaps will have size at least 2. And, indeed, we get all sequences of six numbers without twins this way, and each sequence is produced exactly once. In fact, for each sequence without twins we can tell which sequence in $1, 2, \ldots 43, 44$ it came from. To do this, we just subtract the numbers $0, 1, 2, 3, 4, 5$ from each of them in turn. For example, $6, 8, 31, 37, 40, 48$ arises from the numbers $6, 7, 29, 34, 36, 43$. So, there are $\binom{49}{6} = 13,983,816$ possible lottery results with six numbers chosen from 49, but only $\binom{44}{6} = 7,059,052$ of them have no twins. And (here comes the calculator),

$$\frac{7,059,052}{13,983,816} = 0.5048$$

quite precisely. Thus there's a 50.48% chance of 'no twins'. Good enough for me to bet on, provided I get the chance to do it again and again. Indeed, out of one thousand games I would expect to win 505 that way, that is, ten more than you! In the long run that should pay off!

There are other roughly fifty-fifty games around as well. What is, for example, the chance that the sum of the numbers in the lottery is odd? Of course, that should be roughly fifty-fifty, but it is not exactly one half. Indeed, again a little calculation with binomial coefficients yields that the probability lies at 50.0072%, so it is just a tiny bit more probable that the sum will be even. Do the calculations yourself; it's not that tricky.

Hint: if the sum is to be odd, then either one, or three, or five of the six numbers are odd.

Trust is good, but maths is better. Maths might protect you, for example, from making many stupid bets in the future.

• •

FURTHER READING

[1] Walter Moers (2000). *13.5 lives of Captain Bluebear* (translated by John Brownjohn). Overlook Press.
[2] The UK National Lottery latest results. <http://www.national-lottery.co.uk/player/p/drawHistory.do>.

PYTHAGORAS'S THEOREM: c^2

Hommage à Queneau

Euclid's proof

PROPOSITION 47.

In right-angled triangles the square on the side subtending the right angle is equal to the squares on the sides containing the right angle.

Let ABC be a right-angled triangle having the angle
5 BAC right;
I say that the square on BC is equal to the squares on BA, AC.
For let there be described
on BC the square $BDEC$,
10 and on BA, AC the squares
GB, HC;   [I. 46]
through A let AL be drawn
parallel to either BD or CE,
and let AD, FC be joined.
15   Then, since each of the
angles BAC, BAG is right,
it follows that with a straight
line BA, and at the point A
on it, the two straight lines
20 AC, AG not lying on the
same side make the adjacent
angles equal to two right
angles;
    therefore CA is in a straight line with AG.   [I. 14]
25   For the same reason
        BA is also in a straight line with AH.
    And, since the angle DBC is equal to the angle FBA: for
each is right:
let the angle ABC be added to each:
30   therefore the whole angle DBA is equal to the whole
angle FBC.   [C. N. 2]

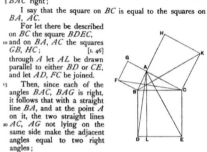

And, since DB is equal to BC, and FB to BA,
the two sides AB, BD are equal to the two sides FB, BC
respectively ,
35   and the angle ABD is equal to the angle FBC:
        therefore the base AD is equal to the base FC,
    and the triangle ABD is equal to the triangle FBC. [I. 4]
    Now the parallelogram BL is double of the triangle ABD,
for they have the same base BD and are in the same parallels
40 BD, AL.   [I. 41]
    And the square GB is double of the triangle FBC,
for they again have the same base FB and are in the same
parallels FB, GC.   [I. 41]
    [But the doubles of equals are equal to one another.]
45   Therefore the parallelogram BL is also equal to the
        square GB.
    Similarly, if AE, BK be joined,
the parallelogram CL can also be proved equal to the square
HC:
50   therefore the whole square $BDEC$ is equal to the two
        squares GB, HC.   [C. N. 2]
    And the square $BDEC$ is described on BC,
    and the squares GB, HC on BA, AC.
    Therefore the square on the side BC is equal to the
55 squares on the sides BA, AC.
    Therefore etc.   Q. E. D.

Reproduced from The Thirteen Books of The Elements:
Volume 1: Books 1 and 2, 2nd edn., trans. Sir Thomas L.
Heath, © 1956, Dover Publications.

Jane Austen

It is a proof universally acknowledged that the square of the hypotenuse must be the sum of the squares of the two shorter sides.

Tweet

OMG, for right angled triangle squares on sides add up :) #pythagoras

Up-goer five*

How long the longest side is of a drawing with three straight sides can be got from how long the other two sides are if one of the two short sides points up and the other points left (or right). How you work this out is to times how long each short side is by how long that same side is and add these two numbers together. Then you have to take the answer and work out what number when taken times that same number gives you this answer. That number you get will be how long the longest side is.

*<http://splasho.com/upgoer5>.

Abraham Lincoln
One-hundred and twenty-four score and seven years ago, our Greek father brought forth on this Earth a new theorem, conceived in liberty, and dedicated to the proposition that the hypotenuse is created equal to the sum of the squares of the two shorter sides.

Scrabble

```
        P       S
    HYPOTENUSE
        T   H   M
      E H   E
    SQUARE
      U G
      A OPPOSITE
      L R   N   H
      S A       E
        SIDES OF
                R
                E
                M
```

Note: 'Q' is centred on the board for best use of double word score.

James Joyce
Yes I love flowers and Stephen likes Pythagoras told me about him and his theorem all about the triangle with its straight sides the right angle in one corner and their names the catheti that touch the right angle and the hypotenuse and how the squares are computed and how if done correctly the squares on the catheti would add up to the hypotenuses square he likes that Stephen when things added up so neatly unlike religion which never added up for him and

roses do they add up I think he would say so he told me how Pythagoras is there all round us in the flowers in the streams and rivers told me how the masons use it make right angles and how it was the secret behind Caesars engines of war and the artificers used it to construct things of length that are not rational how its application leads to spirals and patterns like those in nature I think he sees it everywhere in the trees and in the flowers Yes.

Mills and Boon
She wanted to avoid the love triangle. But what could she do? Taking one more glance at Pythagoras, his sturdy Greek frame, she knew he was Mr Right. He had her number, it all added up. But, what was it? a feeling of unease? he just seemed too wholesome – predictable, almost – square.

Or maybe she should follow her heart and commit to Pierre de Fermat. Her pulse quickened at the thought. He had higher powers. He was real. She could feel her knees weakening as she turned towards his open arms. But how could she be sure? Could he ever be drawn? Would she be marginalised? Where was the proof?

George Orwell
Newspeak (doublethink): $a^2 + b^2 = c^2$, good; $a^3 + b^3 = c^3$, doubleplusgood.

A.A. Milne
Eeyore was depressed. 'How can I find the length of the 500 Acre Wood if no one will listen to me?' No one listened.

Piglet counted all the trees adjacent to the forest edge and came up with a very BIG number.

Not to be outdone, Tigger, sitting opposite, said 'I can come up with an EVEN BIGGER number.' He shot up and bounced Bounced BOUNCED all the way to Rabbit's house.

Christopher Robin said something about 'high rotten yews' that Eeyore didn't understand.

Pooh stared square-eyed and strode off on one of his Thinking Walks. (Tiddly Pom.)

Runes from Middle-earth

ᛁᛏ ᚠᛁᛉ ᚱᛁᚷᚺᛏᛏᛁᚷᛚᛉᛗ ᛏᚱᛁᛁᛁᚷᛚᛗ ᛏᚺᛗ ᚠᚱᛗᚨ ᛒᚠ ᛏᚺᛗ ᛋᛈᚢᚨᚱᛗ ᚠᚺᛗᛋᛗ ᛋᛁᛉᛗ ᛁᛁ ᛏᚺᛗ ᚺᛒᛒᚠᛏᛗᛁᛋᛗ
ᛁᛁ ᛗᛩᚢᚨᛚ ᛏᛈ ᛏᚺᛗ ᛋᛁᛉ ᛒᚠ ᛏᚺᛗ ᚠᚱᛗᛋ ᛒᚠ ᛏᚺᛗ ᛋᛈᚢᚨᚱᛋ ᚠᚺᛗᛋᛋ ᛋᛁᛗᛋ ᚠᚱᛗ ᛏᚺᛗ ᛏᚠᛒ ᛚᛗᚷᛋ.

Begins: 'In any rightangled triangle ...'

50 VISIONS OF MATHEMATICS: FIGURE DESCRIPTIONS AND ATTRIBUTIONS

1. Close-up of a daisy showing Fibonacci spirals. Photograph taken by John Adam, Old Dominion University.

2. A much-magnified small detail of the fractal Mandelbrot set, revealing what appears to be a procession of elephants. Image produced by Philip Dawid, using the program winCIG Chaos Image Generator developed by Thomas Hvel, © Darwin College, University of Cambridge.

3. The Rhind Mathematical Papyrus from Thebes, Egypt, from around 1550 BC contains 84 problems concerned with numerical operations, practical problem-solving, and geometrical shapes. © British Museum.

4. A much-magnified small detail of the fractal Mandelbrot set, containing a scaled replica of the entire set (in solid black) at its centre. Image produced by Philip Dawid, using the program winCIG Chaos Image Generator developed by Thomas Hvel, © Darwin College, University of Cambridge.

5. *Fractal meanders*, an aerial shot taken over northern Florida. Photograph taken by John Adam, Old Dominion University.

6. This 3-dimensional heart shape is an example of an algebraic surface, as defined by the equation below it. Created by Michael Croucher using Mathematica™, first published at <www.demonstrations.wolfram.com/EquationsForValentines/>, © Wolfram Demonstrations Project & Contributors 2013.

7. Viewing the curved surface from three nearby directions, perpendicular to the three planes, gives the three represented apparent contours; the change in these contours is called a beak-to-beak transition. Created by Peter Giblin and Gordon Fletcher. © Peter Giblin.

8. A dramatic shadow of π. Designed by Andrew Martin, <www.smudgeandscribble.co.uk> © Andy Martin & Thumbnail Designs 2007.

9. A solution of the Skyrme model (see Chapter 10) with four nucleons, discovered by Braaten *et al.*, *Physics Letters B*, vol. 235 (1990), pp. 147–152. © Dankrad Fiest.

10. *Bat country*, a 22 feet (6.7 metres) tall Sierpinski tetrahedron composed of 384 softball bats, 130 balls, and a couple of thousand pounds of steel. Designed by Gwen Fisher, engineered by Paul Brown. © Gwen Fisher 2008, <www.beadinfinitum.com>.

11. Illustration of the solution of a sparse linear system of equations using an LU decomposition; black and red indicate the non-zeros and pivots, cyan indicates the columns used, and yellow the consequent build-up of fill-in. © Julian Hall, University of Edinburgh, 2006.

12. The complex folding patterns that arise when a layered material (paper) is put into a test machine and squashed. Created by Timothy Dodwell and Andrew Rhead of the Composite Research Unit, University of Bath.

13. With the advent of 3D printing technology, mathematical constructs such as this quaternion Julia set fractal (upper plot) can be 'printed' as physical objects, in this case in gold-plated brass (lower plot). © Rob Hocking, University of Cambridge 2013.

14. *The proof is in the pudding*, cake illustration of Pythagoras's Theorem, baked by Emiko Dupont for The Great Maths Red Nose Cake Bake Off for Comic Relief at the University of Bath, 2013.

15. Seismic time slice from the North Sea recorded by a 3D seismic survey vessel; the area has undergone a complex tectonic history and exhibits a network of dense polygonal faults distorted into a radial pattern around a salt diapir. © CGG, <www.cgg.com>.

16. A simple BASIC program was written to visualise the digits of π to demonstrate that the sequence is essentially without repeats or structure. Created by Mick Joyce.

17. The Brillouin zones of a square crystal lattice in two dimensions, which underlie the analysis of waves propagating through the crystal. Created by R. R. Hogan, University of Cambridge.

18. Long-exposure photograph of a double pendulum exhibiting chaotic motion. © 2013 Michael G. Devereux.

19. Map showing areas accessible via public transport, visualised in time bands of ten minutes with a departure time of 9.00 a.m. Created by Mapumental using Open Street Map and the National Public Transport Data Repository. Creative Commons license CC-BY-SA 2011.

20. A hyperoval within the Hjelmslev plane over the integers modulo 4. © Michael Kiermaier.

21. Details from the ceiling of the Sagrada Familia basilica in Barcelona, illustrating architect Gaudí's love of mathematical design. © Tim Jones.

22. A physical experiment showing the ascent of a dense vortex formed by the ejection of saline solution (stained blue) upwards into a quiescent fresh-water environment. Image by Ole Myrtroeen and Gary Hunt, University of Cambridge, from the cover of vol. 657 of *Journal of Fluid Mechanics*, © Cambridge University Press.

23. Fluid particle trajectories in a rotating plane-layer convective flow, heated from below; colour corresponds to the time spent in the flow, the darker the colour the longer. Image created by Benjamin Favier, University of Cambridge, using VAPOR (<www.vapor.ucar.edu>) from The US National Center for Atmospheric Research.

24. A surface with many singularities, constructed using computer algebra. Created by Oliver Labs, using the software Singular and visualised using Surf. © Oliver Labs, <www.MO-Labs.com>.

25. Visualisation of a topologically non-trivial solution to the Einstein equations: a wormhole from the Bow River in Calgary, Canada, to St John's College, Cambridge. © Rob Hocking, University of Cambridge 2013.

26. The Queen Elizabeth II Great Court is the largest covered public square in Europe, enclosed under a spectacular glass and steel roof. © British Museum.

27. Magnetic survey data giving the depth and thickness of sedimentary sequences with superimposed surface elevation and seafloor bathymetry. © CGG, <www.cgg.com>.

28. The 'Mandelbox' is a 3D fractal object that represents the points in space that do not move to infinity under the action of a set of geometric transformations. Image by Jos Leys, <www.josleys.com>.

29. A trefoil knot combining four parallel Möbius strips and a spiral tube running continuously round. Drawn freehand by graphic artist Tom Holliday, inspired by M. C. Escher.

30. A minimal energy configuration for 2790 interacting particles on a sphere, computed using a finite element discretisation of a local dynamic density functional theory approximation. Method described in Backofen et al., *Multiscale Modeling & Simulation*, vol. 9, pp. 314–334, 2011. © Alex Voigt.

31. A ball filled with hyperbolic dodecahedrons; five dodecahedrons meet at each edge. Image by Jos Leys, <www.josleys.com>.

32. A mathematical representation using the golden ratio (0.618 …) of 3000 seeds on a large sunflower with various spirals occurring a Fibonacci number of times (1, 1, 2, 3, 5, 8, 13 …). © Ron Knott.

33. For a fixed diameter of string, an *ideal knot* is the configuration with the smallest total length. The image shows a link of two interlocking strings of half their full diameter, exhibiting spontaneous symmetry breaking. © Ben Laurie.

34. A 3D print of a shadow of the tesseract, a 4D version of the cube. Image by Henry Segerman of a sculpture made by Saul Schleimer and Henry Segerman.

35. 'Codex Processianus' created using a 2D domain transformation technique where the polar form of a complex function controls drawing direction and speed of a virtual pen. Created by Martin Schneider using the software available at <www.openprocessing.org/sketch/7579>.

36. Willmore tori equivariant under torus knot actions in the 3-sphere. Created by Nicholas Schmidt, Tübingen University, Germany.

37. A geometric representation of the aberrations seen by a new robust optimisation method; the smaller the shape, the more robust the solution. © Sebastian Stiller.

38. A representation of branching Brownian motion in which each particle moves randomly and, at certain random instances of time, splits into two particles that evolve in the same manner. © Bati Sengul 2013.

39. A Penrose tiling pattern on the floor of the Molecular and Chemical Sciences Building of the University of Western Australia. © Andrew Wrigley.

40. The potentially chaotic motion of a double pendulum revealed using a stroboscope. © Karin Mora, University of Bath.

41. Crocheted model of a hyperbolic plane, a surface with negative curvature. © Created and photographed by Daina Taimina.

42. The limit set of a 3D Kleinian group obtained by the action of three quaternian Möbius transformations. Image by Jos Leys, <www.josleys.com>.

43. Street art showing the equation of the area of a circle. Created by Andrew Hale and Patrick Cully, Research Engineers at Frazer-Nash Consultancy, studying for an EngD in the EPSRC Industrial Doctorate Centre in Systems, University of Bristol.

44. Efficient mathematical methods reveal enzymatic binding processes: the inhibitor molecule be-statin binds to the catalytic centre of aminopeptidase N in a three-step process. Image created by Alexander Bujotzek, Peter Deuflhard, and Marcus Weber. © Zuse Institute Berlin, Germany.

45. A demonstration of the mathematical principles of the original Forth Bridge in Scotland performed at Imperial College in 1887; the central 'weight' is Kaichi Watanabe, one of the first Japanese engineers to study in the UK, while Sir John Fowler and Benjamin Baker provide the supports. Photograph courtesy of the Department of Civil and Environmental Engineering, Imperial College London.

46. Visualisation of a complex-variable function using a method called domain colouring, developed by Frank Farris. Created by Maksim Zhuk using the Pygame library within Python.

47. A hybrid 3D fractal object, obtained using a combination of two sets of geometric transformations. Image by Jos Leys, <www.josleys.com>.

48. Tracers shown as streaks in a plane turbulent water jet confined between two closely separated walls. Image by Julian Landel, Colm Caulfield, and Andy Woods from *Journal of Fluid Mechanics*, vol. 692, pp. 347–368. © Cambridge University Press.

49. A version of the Klein bottle, rendered partially transparent so that the form is clear. © Charles Trevelyan.

50. The image shows an adaptive computational mesh, with high resolution in the regions depicted by the number 50. © Emily Jane Walsh, Simon Fraser University Canada.

LIST OF CONTRIBUTORS

David Acheson is Emeritus Fellow of Jesus College, University of Oxford, UK.

Alan J. Aw likes mathematics and enjoys convincing others to like it. Apart from pursuing nerdish goals, he enjoys evening runs (in sunny hot Singapore) and plays classical or jazz piano music.

John D. Barrow FRS is Professor of Mathematical Sciences and Director of the Millennium Mathematics Project at the University of Cambridge, UK.

Greg Bason is a mathematics lecturer at Abingdon and Witney College in Oxfordshire, UK.

David Berman is Reader in Theoretical Physics at Queen Mary, University of London, UK.

Ken Bray is a theoretical physicist and a Senior Visiting Fellow in the Department of Mechanical Engineering at the University of Bath, UK.

Ellen Brooks-Pollock is a Research Fellow at the Department of Veterinary Medicine at the University of Cambridge, UK.

Chris Budd is Professor of Mathematics at the University of Bath, and at the Royal Institution of Great Britain, UK.

Alan Champneys is Professor of Applied Nonlinear Mathematics at the University of Bristol, UK.

Carson C. Chow is a Senior Investigator at the Laboratory of Biological Modeling, National Institute of Diabetes and Digestive and Kidney Disorders, National Institutes of Health, USA.

Tony Crilly is a writer whose books include *Arthur Cayley: Mathematician Laureate of the Victorian age*, *Mathematics: The big questions*, and *50 mathematical ideas you really need to know*. He is Emeritus Reader in Mathematical Sciences at Middlesex University, UK.

Graham Divall is an independent consultant forensic scientist with 35 years' experience in the field of bloodstain examination.

Marcus du Sautoy is Simonyi Professor for the Public Understanding of Science at the University of Oxford, UK.

Ken Eames is a Lecturer at the London School of Hygiene and Tropical Medicine, UK.

Richard Elwes is a Lecturer at the University of Leeds, UK, and a freelance mathematical writer.

Alistair Fitt is the Pro Vice-Chancellor for Research & Knowledge Exchange at Oxford Brookes University, UK.

Marianne Freiberger is Editor of *Plus Magazine* (<http://plus.maths.org>), a free online magazine about mathematics aimed at a general audience.

Paul Glendinning is Professor of Mathematics at the University of Manchester, UK.

Julia Gog is Reader in Mathematical Biology at the Department of Applied Mathematics and Theoretical Physics, University of Cambridge, UK.

Alain Goriely is Professor of Mathematical Modelling at the University of Oxford, UK, and the Director of the Oxford Centre for Collaborative Applied Mathematics (OCCAM), UK.

Thilo Gross is Reader in Engineering Mathematics at the University of Bristol, UK.

David Hand is Emeritus Professor of Mathematics at Imperial College London, UK.

Andreas M. Hinz is Professor of Mathematics at the University of Munich (LMU), Germany, with a special interest in applications like the modelling of the Tower of Hanoi.

Philip Holmes works on many aspects (mathematical, pedagogical, and historical) of nonlinear science, dynamics, chaos, and turbulence. He teaches at Princeton University, USA.

Steve Humble (aka Dr Maths) works for the Education Department at Newcastle University, UK.

Lisa Jardine is Professor of Renaissance Studies and Director of the Centre for Humanities Interdisciplinary Research Projects at University College London, UK.

Adam Jasko is an undergraduate mathematics student at Nottingham University, UK.

Tom Körner is Professor of Fourier Analysis at the University of Cambridge, UK.

Adam Kucharski is a Research Fellow in the Department of Infectious Disease Epidemiology at London School of Hygiene and Tropical Medicine, UK.

Mario Livio is an astrophysicist at the Space Telescope Science Institute, in Baltimore, Maryland, USA.

Peter Lynch is Professor of Meteorology in the School of Mathematical Sciences, University College Dublin, Ireland. He blogs at <http://thatsmaths.com>.

Maarten McKubre-Jordens is a Lecturer in Mathematics at the University of Canterbury, New Zealand.

Alexander Masters won the Guardian First Book Award for his biography *Stuart, a life backwards* (2005). His second biography, *The genius in my basement* (2011), is about Simon Norton.

Derek E. Moulton is a University Lecturer in the Mathematical Institute at the University of Oxford, UK.

Yutaka Nishiyama is Professor of Mathematical Sciences at Osaka University of Economics, Japan.

Simon Norton is a group theorist and a world expert on the Monster Group and Monstrous Moonshine, a mathematical object so remarkable and unexpected that he calls it 'the voice of God.'

Colva Roney-Dougal is a Senior Lecturer in Pure Mathematics at the University of St Andrews, UK.

Chris Sangwin is a Senior Lecturer in the Mathematics Education Center, Loughborough University, UK.

Caroline Series is Professor of Mathematics at the University of Warwick, UK.

Simon Singh is a writer, whose books include *Fermat's last theorem* and *The Simpsons and their mathematical secrets*.

David Spiegelhalter is Winton Professor for the Public Understanding of Risk at the University of Cambridge, UK.

Ian Stewart FRS is Professor of Mathematics at the University of Warwick, UK.

Danielle Stretch is an administrator in the Department of Applied Mathematics and Theoretical Physics at the University of Cambridge, UK.

Paul Taylor is a graduate student in Systems Biology at New College, University of Oxford, UK.

Rachel Thomas is Editor of *Plus Magazine* (<http://plus.maths.org>), a free online magazine about mathematics aimed at a general audience.

Vince Vatter is an Assistant Professor in Mathematics at the University of Florida, USA.

Ahmer Wadee is Reader in Nonlinear Mechanics at Imperial College London, UK.

Paul Williams is a Royal Society University Research Fellow at the University of Reading, UK.

Eddie Wilson is Professor of Intelligent Transport Systems at the University of Bristol, UK.

Phil Wilson is a Senior Lecturer in Mathematics at the University of Canterbury, New Zealand.

Thomas Woolley is a post doctoral researcher in the Mathematical Institute at the University of Oxford, UK.

Andrew Wrigley is a Senior Teacher of Mathematics at Somerset College, Queensland, Australia.

Günter M. Ziegler is Professor of Mathematics at Freie Universität Berlin, Germany.